Elements of Chemical Engineering

Elements of Chemical Engineering

Madelyn Jackson

MURPHY & MOORE
www.murphy-moorepublishing.com

Elements of Chemical Engineering
Madelyn Jackson
ISBN: 978-1-63987-186-5 (Hardback)

MURPHY & MOORE
Published by Murphy & Moore Publishing,
1 Rockefeller Plaza,
New York City, NY 10020, USA

Cataloging-in-Publication Data

Elements of chemical engineering / Madelyn Jackson.
p. cm.
Includes bibliographical references and index.
ISBN 978-1-63987-186-5
1. Chemical engineering. 2. Chemistry, Technical. 3. Engineering. I. Jackson, Madelyn.
TP155 .E44 2022
660--dc23

For more information regarding Murphy & Moore Publishing and its products, please visit the publisher's website www.murphy-moorepublishing.com

Table of Contents

Table of Contents

Preface

The branch of engineering which focuses on the operational techniques, design processes and methods to increase production efficiency is known as chemical engineering. It draws on the principles of physics, mathematics, chemistry, economics and biology for efficient transportation and transformation of energy and materials. Some of the important aspects of chemical engineering are process design and analysis, plant design and construction, and transport phenomenon. It finds extensive applications across a wide variety of fields such as polymers, medical devices, fertilizers, foods, drugs, petrochemicals and microelectronics. Chemical engineering also applies concepts from nanomaterials and nanotechnology to industrial processes responsible for converting raw materials and chemicals into useful products. This book is a compilation of chapters that discuss the most vital concepts in the field of chemical engineering. It elucidates new techniques and their applications in a multidisciplinary approach. As this field is emerging at a rapid pace, the contents of this book will help the readers understand the modern concepts and applications of the subject.

A short introduction to every chapter is written below to provide an overview of the content of the book:

Chapter 1- The branch of engineering which focuses on studying the operation and design of chemical plants is termed as chemical engineering. It is also involved in improving their production. A few of its subtypes are biochemical engineering and electrochemical engineering. This is an introductory chapter which will introduce briefly all these significant aspects of chemical engineering; **Chapter 2**- There are various principles and concepts which are central to the discipline of chemical engineering. Some of these are membrane process, adsorption, crystallization process, nitration, and sulfonation and sulfation. These diverse principles and concepts of chemical engineering have been thoroughly discussed in this chapter; **Chapter 3**- The process to convert a mixture of chemical substances into two or more distinct product mixtures is termed as chemical separation. A few techniques used for chemical separation are chromatography, extraction, filtration and centrifugation. All these chemical separation techniques have been carefully analyzed in this chapter; **Chapter 4**- There are various aspects which are vital to develop an understanding of chemical engineering. A few of these are transport phenomena, chemical thermodynamics, process simulation and chemical kinetics. This chapter closely examines these key concepts of chemical engineering to provide an extensive understanding of the subject; **Chapter 5**- Chemical engineering is used in a wide range of fields and industries such as food engineering, plastics engineering, pharmaceutical engineering, tissue engineering, chemometrics and ceramic engineering. The topics elaborated in this chapter will help in gaining a better perspective about these applications of chemical engineering.

I extend my sincere thanks to the publisher for considering me worthy of this task. Finally, I thank my family for being a source of support and help.

Madelyn Jackson

1

Understanding Chemical Engineering

The branch of engineering which focuses on studying the operation and design of chemical plants is termed as chemical engineering. It is also involved in improving their production. A few of its subtypes are biochemical engineering and electrochemical engineering. This is an introductory chapter which will introduce briefly all these significant aspects of chemical engineering.

Chemical engineering uses mathematics, physics, chemistry and biology to develop chemical transformations and processes to create useful products and materials that improve society. It is one of the most broadly based engineering disciplines, creating high demand for chemical engineers across a variety of industries and professions.

Chemical engineers devise ways to convert raw materials and intermediate chemicals into more useful forms. They develop processes and products that are energy efficient, environmentally friendly and profitable. Chemical engineers make petrochemicals, polymers, microelectronics, biomedical devices, paper, dyes, drugs, fertilizers and foods. Practically every product has been improved by chemical engineering in some way.

Chemical engineers work to solve major societal challenges in energy, human health, manufacturing and sustainability. They are employed in the petrochemical, pharmaceutical, microelectronics, biomedical and environmental industries. The breadth and depth of training in chemical engineering allows graduates to go on to graduate programs and careers in academia, national laboratories, medicine business and law. The versatility of chemical engineers allows them to make advances that enhance people's everyday lives across all walks of life.

Biochemical Engineering

Biotechnology is usually taken to mean the use of sophisticated genetic techniques outside the cell to result in the production of useful products. However, many people broaden the term to include applied biology, or even engineering. Strictly speaking, biochemical engineering is usually defined as the extension of chemical engineering principles to systems using a biological catalyst to bring about desired chemical transformations. It is often subdivided into reactor design and downstream separation.

Despite the appearance that it is a new discipline, biochemical engineering as such dates back to the turn of the century, when certain rudimentary principles were used in the biological treatment of wastewater and in the production of acetone and butanol for explosives. Furthermore, fermentation ethanol was one of the basic raw materials for the chemical industry until the Second World War. The modern genesis of the discipline dates back to the early 1940s when penicillin was required in vast quantities.

The recent development of biochemical engineering is the direct result of advances in molecular biology (e.g., recombinant DNA, tissue culture, protoplast fusion, monoclonal antibodies) and protein engineering. Biochemical engineers work in a wide cross-section of industry including; pharmaceuticals, food, fine chemicals, wastewater treatment, mining and energy. Hence, biochemical products vary from high volume, low coast to very low volume and extremely high cost. Figure details this wide variation in cost.

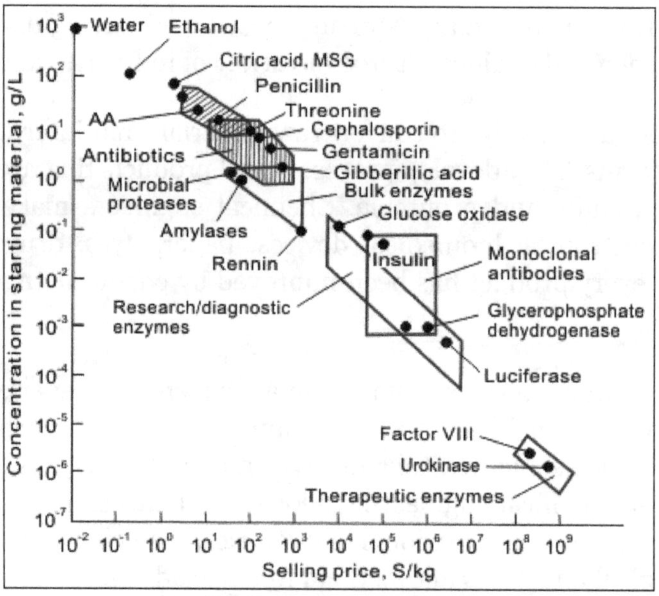

Concentration in the broth versus final selling price.

Biochemical engineering involves the use of a catalyst, either an enzyme or whole cell, which is either freely suspended in an aqueous medium or "immobilized" in a gel or attached to a solid surface. In contrast to an enzyme, whose activity decreases over time, cells are autocatalytic ($dx/dt = \mu x$, where x = cell concentration and μ = growth rate, t^{-1}) and can grow exponentially with excess nutrients. Reactor conditions are typically: a pH of 6–8; a temperature of 15–16 °C; atmospheric pressure; and a cell concentration of 2–40 g/l. Reactors can be operated in either batch mode (most common in the pharmaceutical industry), typically over 24–48 hours, or continuously (e.g., wastewater treatment). Reactors commonly used vary from CSTRs (continually-stirred tank reactors), through plug flow (whose efficiency is similar to CSTRs due to back mixing), to specialized reactors which immobilize the cells or enzymes to prevent them from being washed out of the reactor due to their small size and density (SG = 1.04). These types of reactors essentially "de-link" the hydraulic retention time from the cell retention time, thereby enabling process intensification to occur.

The downstream separation of biological products is one of the major challenges in biochemical engineering. These products are often present in very low concentrations (100 mg/l), and are extremely labile to mechanical shear and extremes in pH and temperature. The difficulty of the separation process depends very strongly on the product itself and whether it is the biomass, an extracellular product excreted by the cell, or an intracellular protein or inclusion body. Since most fermentation broths are extremely complex mixtures of a variety of organic and inorganic constituents and most products have to meet extremely stringent standards in terms of purity especially the therapeutics — recovery and concentration can often involve 10–15 unit operations. This can lead to quite low overall recoveries (10–20%); hence, separation can constitute as much as 70% of the overall product cost. The separation techniques used include those based on size, diffusivity, charge, surface activity, density and polarity, and often include techniques rarely seen in the chemical process industry.

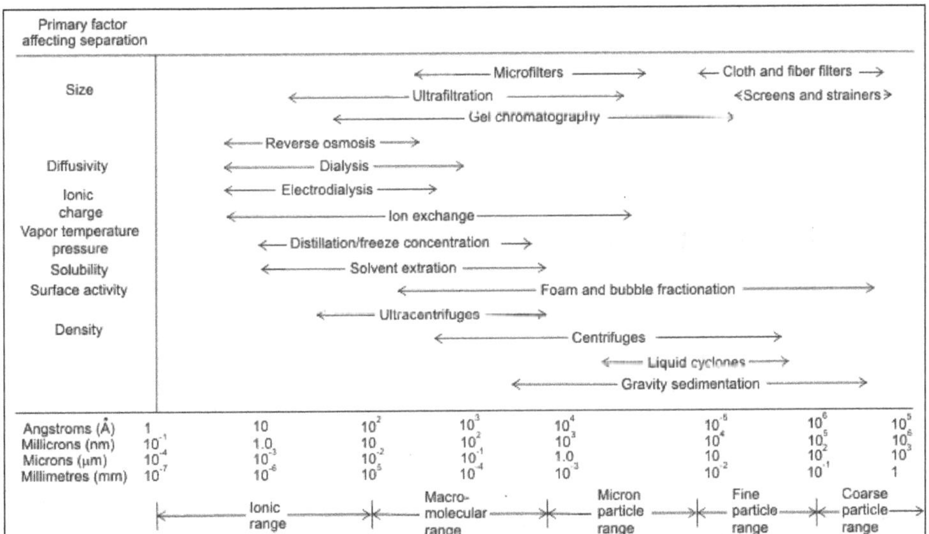

Ranges of applications of various unit operations.

The modeling of most biochemical engineering processes is still fraught with difficulties, although simple enzymatic reactions can be represented by the theoretically derived Michaelis-Menten relationship.

$$V = \frac{V_{max} c}{K_m + c}$$

where,

- V = rate of reaction, t^{-1};

- V_{max} = maximum rate of reaction, t^{-1};
- K_m = Michaelis-Menten, or half rate constant, ML^{-3};
- c = concentration of substrate, ML^{-3}.

However, the modeling of whole cell reactions is far more complex since cell metabolism consists of many sequential reactions (10^3–10^4), any one of which may be rate limiting at any point in time. In addition, the release of product may be kinetically associated with cell growth ("growth associated"), or occur after growth has virtually ceased ("secondary metabolite"), or be a complex mixture of both. In modeling growth kinetics, it is possible to take two approaches: using an "unstructured" model, which takes an uncritical "black box" stance; or developing a "structured" model, which is based more on the basics of cell metabolism. Due to its simplicity and relative success, the most commonly-used cell growth model is the unstructured one developed empirically by Monod:

$$\mu = \frac{\mu_{max} c}{K_s + c}$$

where

- μ = rate of cell growth, t^{-1};
- μ_{max} = maximum rate of cell growth, t^{-1};
- K_s = half rate constant, ML^{-3};
- c = concentration of rate limiting substrate, ML^{-3}.

While empirically derived, the equation has obvious similarities to the Michaelis-Menten equation for enzyme kinetics.

Electrochemical Engineering

ECE is the branch of chemical engineering (ChE) concerned with the conception, design, and operation of electrochemical systems. Although the subject of electrochemistry has been known and studied for over 150 years, as a full-fledged wing of ChE, ECE came of age in the forties and fifties when the concept was realized that any electrode reaction, despite its characteristics, is amenable to the same kind of fundamental approach as is applied to heterogeneous catalysis. The second concept that delineated ECE as a branch of ChE is that the rate of electrode reactions is generally affected by the catalytic properties of the substrate of the electronic conductor as well as by the interplay of mass transport mechanisms that determine the diffusion rate of electroactive species.

Electrochemical Reaction Engineering

Electrochemical reaction engineering is an area which integrates the mechanistic, kinetic and transport behavior of electrode processes. Researcher studied the reactions which take place at the electrodes of electrolytic cells. He discovered the two laws which express the relationships between the amounts of products formed during electrolysis and the quantity of electricity passed. These can be conveniently combined in the following statement: The passage of 96487 coulombs through an electrochemical reactor produces in total one gram equivalent of products at an electrode. This quantity of electricity is called the Faraday and is designated by F. In the case of a number of reactions occurring at an electrode with n_j moles of j produced by a current I_j flowing for t seconds we can write,

$$\sum n_j = \frac{1}{F}\sum \frac{1_j}{z_j}$$

where z_j is the number of electrons associated with the production of a molecule of j, $\sum n_j$ is the total number of moles produced. I_j refers to the partial current associated with the production of a given species. The total current I is evidently the sum of partial current.

Just as in chemical catalysis, electrocatalysis provides a reaction path which lowers the energy of activation and hence increases the rate of reaction, i.e., the current density for a given overvoltage. We can distinguish between the situation when a species in solution acts as a catalyst and a process where the reactive intermediate has to be adsorbed on the electrode. The epoxidation of propylene by means of bromide ions is a typical analysis. Industrial Electrochemistry, demonstrates reasoning that can sometimes lead to a decision on the most probable reaction path when an adsorption step is involved. Electrocatalysis is of major importance industrially. Use of dimensionally stable anodes in the chlorine industry is a good example.

Electrode Processes and Reaction Velocity

Based on the global concepts of heterogeneous chemical kinetics the difference between the rates of the forward (cathodic) and reverse (anodic) reactions of the general electrode reaction, $A + ze^- \leftrightarrow B$ at equilibrium can be written as,

$$i = zFK_{f_1}C_{A_S} - zFk_{b_1}C_{B_S}$$

By using an Arrhenius type of rate constant/activation energy relationship, the rate constants k_{f1} and k_{b1} can be expressed in term of the electrode potential, which is a measure of the free energy requirements.

$$k_{f_1} = k_{f_1}^{O}\exp\left(\frac{-\alpha zF\left(-V_C^*\right)}{RT}\right) \text{and } k_{b_1} = k_{b_1}^{O}\exp\left(\frac{(1-\alpha)zF\left(-V_C^*\right)}{RT}\right)$$

The above equations imply that a fraction, $\alpha(-V_C^*)$ of the cathode potential is effective in promoting the cathodic process the remainder $(1-\alpha)(-V_C^*)$ suppressing the reverse (anodic) process. At equilibrium, i = 0 and the concentrations of A and B at the surface equals to bulk, thus we can easily manipulate the rate equation into the general form:

$$i = i_O \left[\frac{C_{A_S}}{C_{A_B}} \exp\left(\frac{-\alpha zF\eta}{RT}\right) - \frac{C_{B_s}}{C_{B_b}} \exp\left(\frac{(1-\alpha)zF\,\eta}{RT}\right) \right]$$

This is the Butler-Volmer equation. It is a standard model that can be used to describe the current-overpotential relationship for an electrode at a specified temperature, pressure and concentration of reacting species. The overpotential η represents the departure from the equilibrium potential. i_o is the exchange current density represents the rates at equilibrium. An attempt to introduce some additional terms into B-V equation to allow elucidation of reaction mechanisms from current density/overpotential measurements is provided by the concept of an electrochemical reaction order due to Vetter. A plot of the current-overpotential relation is commonly referred to as a polarization curve. For certain reactions B-V equation provides a reasonable description of current-overpotential behavior. Even when the reaction mechanism is know it may not be possible to cast the overall current-overpotential relation in B-V form; in those cases an alternative simplified equation can often be developed. In the special case where the transfer coefficients are equal, the exponentials can be cast in the form of a hyperbolic sine function. When the magnitude of the overpotential is large, another simplification is possible. If the overpotential is large and positive, then the first exponential term is much larger than the second term and the polarization expression becomes,

$$\eta = \frac{RT}{\alpha zF}(\text{In } i_o - \text{In } i)$$

This expression is the Tafel equation. It is a good approximation to B-V equation for overpotential of more than 100 mV in absolute magnitude. Polarization has three underlying causes: ohmic, activation, and concentration. Their effects are more or less addition and the potential shift is called an overpotential η_{tot}.

$$E = E_{eq} + \eta_{tot}$$

$$\eta_{tot} = \eta_{ohmic} + \eta_{act} + \eta_{Conc}$$

The parameters that define the rate of a chemical reaction are determined generally by the broad area of the reaction that is, chemical, photochemical, electrochemical, etc. In electrochemical reactions, the characteristic measurable parameter is the current density i, which is the current in A/m² of the electrode, and the parameter that converts it to molar rate is the Faraday. Thus the reaction rate (for the disappearance of A) if the case of cathodic reduction alone is considered can be expressed as,

$$-r_A = \frac{V_r}{A_C}\frac{dC_A}{dt} = \frac{i_A}{zF}$$

Alternatively the rate can be written as,

$$\frac{i_A}{zF} = k_{f_1}\left(C_{A_s}\right)^m$$

where, m is the order of reaction. The rate of reaction at an electrode surface is also governed by mass transport, defined through the use of a mass transfer coefficient k_{LA} for diffusing species A as,

$$i_A = zFk_{LA}\left[C_A - C_{AS}\right].$$

Eliminating surface concentration from the above equations the electrochemical reaction rate expression for a first order process becomes,

$$\frac{i_A}{zF} = \frac{k_{f_1} C_A}{1 + \frac{k_{f_1}}{k_{LA}}}$$

Following the same procedure the reaction rate (assuming first order kinetics) for the general redox reaction after eliminating the surface concentrations of A and B can be expressed as,

where, $k_{f_1} = k^o_{f1} e^{-\alpha_1 fE}$

$$k_{b1} = k^o_{b1} e^{-\beta_1 fE}$$

This is a general and useful way of expressing electrochemical reaction rate of any redox reaction at an electrode.

Complex Electrochemical Reaction Schemes

Mathematical models of complex electrochemical reaction schemes are developed in which mass transport of reactants and products at the electrode surface plays an important role in describing overall reaction rates. Cell operation is restricted to potentiostatic operation which allows the model solutions to be obtained in analytical form. The model can be used to simulate the behavior of a potentially useful organic synthesis like the electroreduction of oxalic acid to glyoxylic acid. Consider the reaction scheme for a series of reactions,

$$A + 2e^- \underset{kb_1}{\overset{kf_1}{\rightleftharpoons}} B + 2e^- \underset{kb_2}{\overset{kf_2}{\rightleftharpoons}} E$$

The governing reaction rate equations in terms of kinetic and mass transfer parameters can be written in dimensionless form as,

$$\dot{x}^b_A = -k^*_{mA}\left(x^b_A - x^s_A\right)$$

$$\dot{x}_A^s = -k_{mA}^*\left(x_A^b - x_A^s\right) - k_{f_1}^* x_A^s + k_{b_1}^* x_B^s$$
$$\dot{x}_B^b = k_{mB}^*\left(X_B^s - X_B^b\right)$$
$$\dot{x}_B^s = -k_{mB}^*\left(X_B^s - X_B^b\right) + k_{f_1}^* X_A^s - k_{b_1}^* x_B^s - k_{f_2}^* x_B^s + k_{b_2}^* x_E^s$$
$$\dot{x}_E^b = k_{mE}^*\left(X_E^s - X_E^b\right)$$
$$\dot{x}_E^s = -k_{mE}^*\left(X_E^s - X_E^b\right) + k_{f_2}^* X_B^s - k_{b_2}^* X_E^s$$

where $x_i = C_i / C^b{}_{Ao}$ is the dimensionless concentration of species $i, k_{fi}^* = ak_{fi}t_f$ and $k_{bi}^* = ak_{bi}t_f$ for i = 1 or 2, $\dot{X}_i = dx_i / dt^*$ is the dimensionless reaction velocity, and $t^* = t / t_f$ is the dimensionless time. The initial conditions become,

$$x_A^b(0) = 1.0$$
$$x_i^b(0) = 0 \text{ where } i = B \& E$$
$$x_i^s(0) = 0 \text{ where } i = A, B \& E$$

By combining these equations to eliminate the surface concentrations, the reaction rates for both steps become,

$$\frac{i_1}{n_1F} = x_1C_A^b + x_2C_B^b + x_3C_E^b$$
$$\frac{i_2}{n_2F} = y_1C_A^b + y_2C_B^b + y_3C_E^b \text{ and } i_T = C_{dl}\frac{dE}{dt} + i_1 + i_2$$

where x_i and y_i (i = 1, 2, 3) are complex non-linear functions of the electrode potential, rate constants and liquid/solid mass transfer coefficients. Similarly by considering the reaction scheme for the parallel reactions, the reaction rates can be written.

One of the most striking questions of industrial importance is; what relationships exist between current efficiency and product yield? In view of the fact that any industrial process as a first step will correlate profit based on input and output. Thus the question may be answered in a further practical manner as follows:

- No parallel reactions: current efficiency and product yield equal to 1.
- Electrochemical parallel reaction of reactant alone: current efficiency and product yield less than 1, but equal to each other.
- Chemical parallel reaction of reactant alone: current efficiency equals 1, product yield less than 1.
- Electrochemical and chemical parallel reactions of reactant: current efficiency and product yield less than 1, product yield is less than current efficiency.

- Electrochemical parallel reaction of another reactant alone: current efficiency less than 1, product yield equals 1.
- Electrochemical parallel reaction of the reactant and another reactant: current efficiency and product yield less than 1. Product yield is greater than current efficiency.
- Electrochemical and chemical parallel reaction of reactant, coupled with electrochemical parallel reaction of another reactant: current efficiency and product less than 1.

Electrochemical Reaction followed by Series or Complex Chemical Reactions

The coupled chemical-electrochemical reaction sequence is similar to that for the reduction of nitrobenzene (A) to the desired product p-aminophenol (D) and the undesired product aniline (U) through the intermediate phenylhydroxylamine (I). Here the reactant A is electrochemically reduced to a stable intermediate I which is itself a reactant for two competing parallel reactions: a homogeneous chemical decomposition to the desired product D, or a further electrochemical reduction to an undesired product U, $A \xrightarrow{E} I \xrightarrow{C} D, I \xrightarrow{E} U$. Taking into account the dynamic concentration changes at the electrode surface, the reaction rates are,

$$\frac{dC_A^b}{dt} = -k_{mA}\left(C_A^b - C_A^s\right)$$

$$\frac{dC_A^s}{dt} = k_{mA}a\left(C_A^b - C_A^s\right) - k_{1_0}ae^{-\alpha_1 fE}C_A^s$$

$$\frac{dC_I^b}{dt} = -k_{mI}\,a\left(C_I^s - C_I^b\right) - k_3C_I^b$$

$$\frac{dC_I^s}{dt} = -k_{mI}\,a\left(C_I^s - C_I^b\right) + k_{1_0}ae^{-\alpha_1 fE}C_A^s - K_{2_0}ae^{-\alpha_2 fE}C_I^s$$

$$\frac{dC_U^b}{dt} = k_{mU}\,a\left(C_U^s - C_U^b\right)$$

$$\frac{dC_U^s}{dt} = -k_{mU}\,a\left(C_U^s - C_U^b\right) + k_{2_0}ae^{-\alpha_2 fE}C_I^s$$

$$\frac{dC_D^b}{dt} = k_3C_I^b$$

Here C_i is the bulk concentration of the species i, C_i^s, is the concentration at the electrode surface, which differs form the bulk concentration C_i^b because of the presence of mass transfer resistance, k_{mi} is the mass transfer coefficient of species i, f = F/RT, α is the transfer coefficient and a is the specific electrode area. Only reactant A is present initially, which results in the initial conditions:

$$C_A^b(0) = C_{A_0}^b$$

$$C_i^b(0) = 0 \quad \text{where } i = I, U \& D$$

$$C_i^s(0) = 0 \quad \text{where } i = A, I \& U$$

The analytical solutions of the above ODEs using the boundary conditions are given in at steady state. The expressions obtained are suitable for use in reactor design, modeling and control. The other complex multiple reactions initiated by a charge transfer step at an electrode surface is anodic formation of chlorate and the epoxidation of propylene. For epoxidation of propylene, application of concentration profiling enables an accurate and straightforward analytical solution to be obtained.

A well-founded model of electrochemical engineering apparatus should be based on the equations of change for mass, energy and momentum. At present, the derivation of a model from the basic equation is an error prone and time consuming task normally requiring specific expertise. At the same time simulation techniques are gaining more importance in industry. This leads to an increasing gap between supply and demand of well-defined, consistent models. The expert system (ES) methodology commonly used in many engineering practice today give enough support for the modeler in the task of deriving a specific model. Extending the model library with electrochemical reaction models is also possible, but studies report only very few application specific ES.

We emphasize gain the need for reaction modeling. Because the plot of product formation as a function of mass transfer rate enables us to specify the required mass transfer characteristics of any cell intended to be used in a synthesis plant. A relatively simple model (if it is appropriate; some uncertainty in the numerical values of the model constants is not usually important) enables us to select the variables that govern performance. Experimentation, particularly at the expensive pilot plant stage, can be minimized and development costs significantly reduced. Development of a reaction model is an essential first step to obtaining a reactor model, which in turn is a must for process optimization.

Electrochemical Reactor Engineering

In electrochemical terminology, the vessel in which electrochemical transformation occurs is called the cell. To bring the terminology in line with that of chemical reaction engineering, we refer to it as an ECR. An ECR is a device in which one or more chemical species are transformed to alternative states with an associated energy change. The electrochemical reactor unit can be made up of many cells which are suitably connected together both electrically and hydraulically. Factors that are important in the electrochemical reactor design are: productivity, energy requirement, cell voltage, temperature control, hydrodynamics, mass transport, reactor operation factors, electrode, membrane and other materials. The form a reactor takes depends largely on the number of phases involved and the energy requirements. The size depends on the rate of reaction. It is the unique feature of ECR that the current and hence the reaction rate can readily be measured. The flux N of any species j across a plane of cross-section area A is given by,

$$N_j = D_j \nabla C_j - \frac{z_j F}{RT} D_j C_j \nabla \phi + C_j u$$

The bold-faced quantities in the above equation signify that they are vectorial; i.e., they have both magnitude and direction. The flux has the unit of mol m^{-2} s^{-1}. D is the diffusion coefficient m^2/s, C_j is the species concentration mol/m3. $\nabla\varphi$ is the potential gradient (i.e., the electric field V/m), z_j is the ionic charge, u is the solution velocity, and F, R and T have their usual significance. We see that the overall flux and mass transport is driven by three terms: diffusion, migration and convection. The driving force for each of the three components is concentration, potential and density gradients respectively. Further, convection can be either natural or forded in an electrochemical system. Two other concepts related to the flow of charge and current flow are electrical conductance and mobilities of both ionic and electronic species. For ECR, in general, the cell and electrode arrangements are so complex that specification of the velocity and concentration variations with time and location cannot be made. Thus, the requirement for the simultaneous solution of the convective mass transfer equation and the equation o of motion (momentum transfer), even in the Navier-Stokes format (constant density and viscosity), cannot be met except in a few cases.

Mass transfer analysis can be considered to proceed through one of three approaches. From theoretical or empirical analyses extensive complications of data on mass transfer have been correlated. These correlations usually employ dimensionless relationships from which mass transfer coefficients can be determined for the given geometric configurations and flow conditions. Some of the models for mass transfer coefficients are given in Table. Analogies between momentum and mass transfer have been developed through such approaches. The overall steps of the electrochemical process are shown in Figure. The technical information needed to choose an ECR are given in Table. It can be seen that decision 'B' will result in a greater effort in design, higher costs and also complex construction, however, provide operational flexibility. Although one tries to use undivided cells, diaphragms are sometimes essential. The voltage loss due to their presence can be greater than 10% of the total voltage loss in the cell. The effective conductivity k_d of a diaphragm saturated with electrolyte can be predicted from knowledge of its hydraulic permeability p and its hydraulic radius r_H, for the relationship.

Table: Technical details for the selection of ECR.

Factor	Choice A	Choice B
Mode of operation	Batch	Continuous
Number of anodes and cathodes	Single	Multiple
Electrode geometry	2-dimensional	3-dimensional

Electrode motion	Static	Moving
Electrode connections	Monopolar	Bipolar
Inter-electrode gap	Moderate	Capillary
Electrolyte manifolding	External	Internal
Cell division	Undivided	Divided
Sealing of cell	Open cell	Closed cell
Types of electrolyte	Liquid	Solid polymer

$$\frac{\kappa}{\kappa_d} = \frac{0.272 r_H^2}{p}$$

The foremost task of any engineering activity is the application of scientific knowledge to the obtention of optimum results under the economic standpoint. Since the electricity required per unit of product greatly depends on plant size and design, optimization plays a vital role in electrochemical industries at large. Several optimization techniques have been reported in literature for the least-cost design of multi-product process. Optimization methods coupled with modern tools of computer-aided designs are also being used to enhance creative process of conceptual and detailed design of engineering systems. Genetic algorithms (GAs) are part of evolutionary computing, which is a rapidly growing area of artificial intelligence. Over last decade, GAs have enjoyed a large scale application for wide spectrum of engineering design problems. But it has been seldom applied for optimizing electrochemical reactor. Vijayasekaran and Basha et al are the first one to model ECR by GAs. The authors studied the effect of variables such as: current density, mass transport, reaction velocity, conversion, yield of the multiple reaction, and concluded that GAs found optimum results that satisfies both industrial practice and process economics. These results are encouraging and suggest the use of GAs in similar reactor design problems.

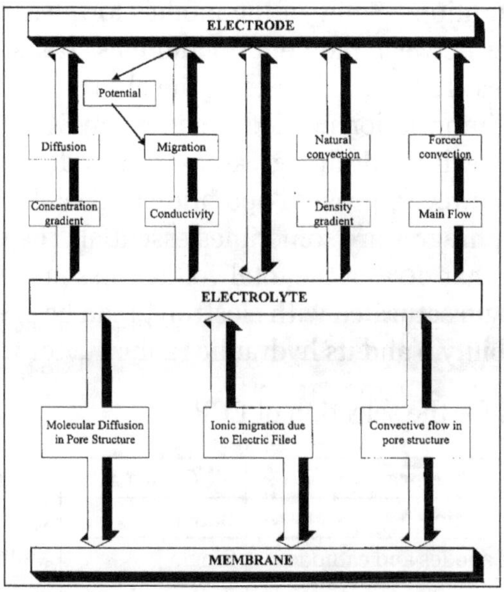

Schematic of ECR phenomenological sequences.

Table: Models for Mass Transfer Coefficients.

Model	Equation	Corresponding Exponents		Description
		Re	Sc	
Film theory	$k=\frac{D}{t}$	–	1.0	Simple, Film thicknesst unknown.
Penetration theory	$k=2\sqrt{\frac{D}{\pi t_1}}$	0.5	0.5	Penetration time t_1 usually unknown.
Surface renewal theory	$k=2\sqrt{\frac{D}{\pi t_1}}$	–	0.5	Renewal rate unknown.
Eddy diffusiveity	$\underline{D \quad E}$	–	–	E_D not calculable.
Boundary layer theory	$k=0.626\frac{D}{L}\left(\frac{L.V}{\gamma}\right)^{1/2}\left(\frac{\gamma}{D}\right)^{1/3}$	0.5	2/3	Derivable simple cases only (e.g. laminar flow past flat plate).

Mechanism of Heat Transfer

Modes of heat transfer (a.k.a. heat transport):

- Conduction
- Convection
- Radiation

Conduction through a solid or a stationary fluid.	Convection from a surface to a moving fluid.	Net radiation heat exchange between two surface.

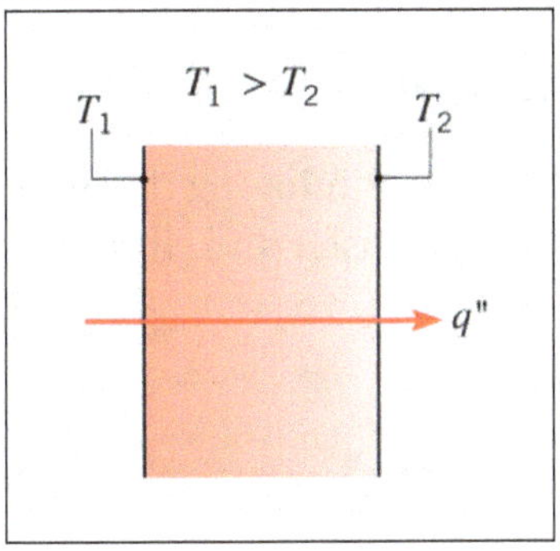

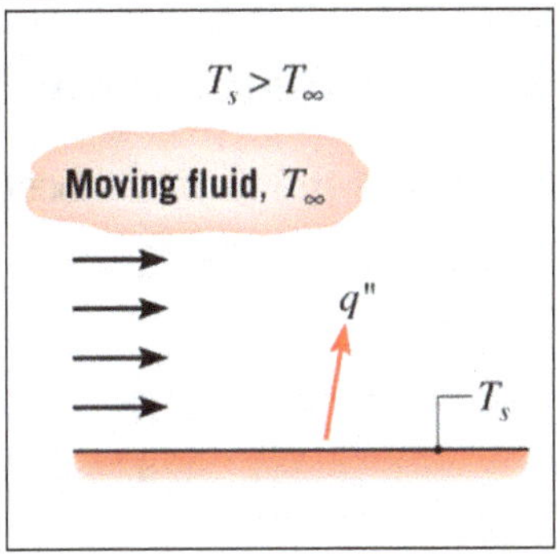

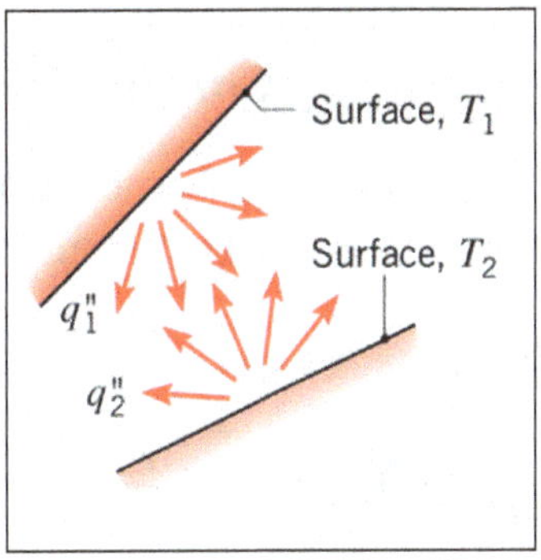

- Conduction: Heat transfer in a solid or a stationary fluid (gas or liquid) due to the random motion of its constituent atoms, molecules and/or electrons.
- Convection: Heat transfer due to the combined influence of bulk (advection) and random motion for fluid flow over a surface.
- Radiation: Energy that is emitted by matter due to changes in the electron configurations of its atoms or molecules and is transported as electromagnetic waves (or photons).

Conduction and convection require the presence of temperature variations in a material medium. Although radiation originates from matter, its transport does not require a material medium and occurs most efficiently in a vacuum.

The heat rate, $\dot{Q}$, is the amount of heat moving through a surface or plane per unit time. Typical units would be J/s. The heat flux, Q/A, is the amount of heat moving through a surface or plane per unit time per unit area. Typical units would be J/sm².

Conduction

Conduction is the heat transfer analogue of diffusion in mass transfer. It is the heat transfer in a solid or a stationary fluid (gas or liquid) due to the random motion of its constituent atoms, molecules and /or electrons.

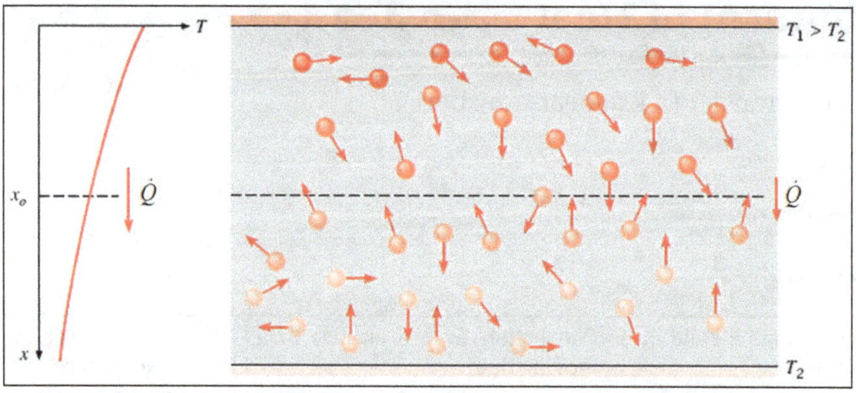

Describing Conduction Heat Transfer Quantitatively

Joseph Fourier: A French mathematician and physicist best known for initiating the investigation of Fourier series and their applications to problems of heat transfer and vibrations.

Fourier's Law

$$\dot{Q}_{cond} = -k\,A\frac{dT}{dx}$$

$$\approx -k\,A\frac{T_2 - T_1}{x_2 - x_1} = -k\,A\frac{\Delta T}{\Delta x}$$

where,

$\dot{Q}_{cond}$	Conduction heat-transfer rate through area A in the x direction between locations 1 and 2.	J/s(or W)
A	Cross-sectional area through which conduction occurs.	M^2
k	The thermal conductivity of the material.	W/mK
T	Temperature.	K

Fourier's law is the heat transfer analogue to Fick's law for diffusion mass transfer.

Thermal Conductivity

Like the diffusion coefficient, D_{AB}, for mass transfer, the thermal conductivity, k, is a material-dependent, constant of proportionality in Fourier's law. Here are some values of the thermal conductivity for various materials:

Material	Thermal Conductivity ($W\ m^{-1}\ K^{-1}$)
Carbon Nanotubes	2000
Diamond	900 – 2320
SiC	490
Silver	424
Copper	398
Gold	315
Aluminum	273
Graphite	119–165
Iron	80
Cupric Oxide	77
Alumina	40
Zinc Oxide	29
Carbon Nanofibers	13
Titania	8.4
Iron (II, III) Oxide	7.0
Bismuth Telluride	5.0
Carbon (amorphous)	1.59
Silica	1.34
C_{60}–C_{70} (Fullerenes)	0.4
Water	0.608
Glycerol	0.285
Ethylene Glycol	0.257
Ethanol	0.172
Decene	0.14

Toluene	0.133
Perfluorotriethylamine	0.13
Hexane	0.126
Poly (α-olefin) oil	0.117
Perfluorohexane (FC-72)	0.057

Thermal conductivity values have even been measured for various biological tissues.

Convection

Convection is the heat transfer analogue of mass transfer convection. Convective heat transfer is heat transfer due to the combined influence of bulk and random motion for fluid flow over a surface. Consider flow across a flat surface.

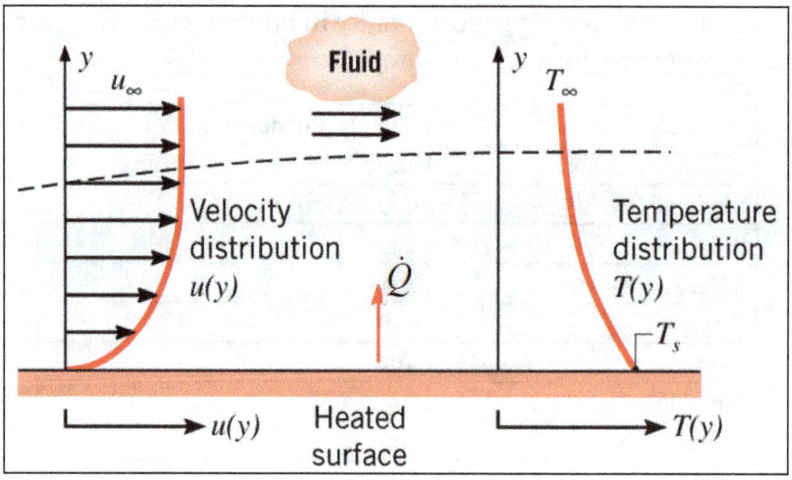

Describing Convective Heat Transfer Quantitatively

Newton's law of cooling:

$$\dot{Q}_{conv} = h\,A\left(T_s - T_{bulk\ fluid}\right) = h\,A\left(T_s - T_\infty\right)$$

where,

$\dot{Q}_{conv}$	Conduction heat-transfer rate through area A in the x direction between locations 1 and 2.	J/s (W)
A	Cross-sectional area through which conduction occurs.	m²
k	The heat transfer coefficient.	W/m² K
T	Temperature.	K

The Heat Transfer Coefficient

Like the mass transfer coefficient, the heat transfer coefficient in strongly influenced by the boundary layer. In particular, it depends on the geometry, flow conditions, fluid, and process. They are not constant for a type of fluid; they depend on the flow conditions, geometry, and fluid properties. Here are some typical values:

Process	h $(W/m^2 \cdot K)$
Free convection	
Gases	2–25
Liquid	50–1000
Forced convection	
Gases	25–250
Liquid	100–20,000
Convection with phase change	
Boiling or condensation	2500–100,000

Radiation

Radiation is energy that is emitted by matter due to changes in the electron configurations of its atoms or molecules and is transported as electromagnetic waves (or photons). There is no mass transfer analogue to radiation. The electromagnetic spectrum and the range of wavelengths involved in radiative heat transfer.

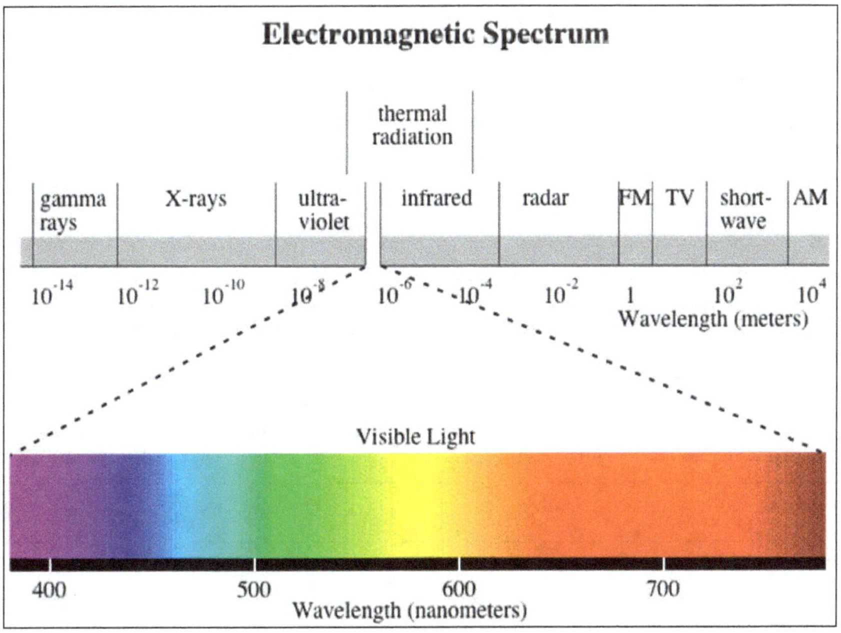

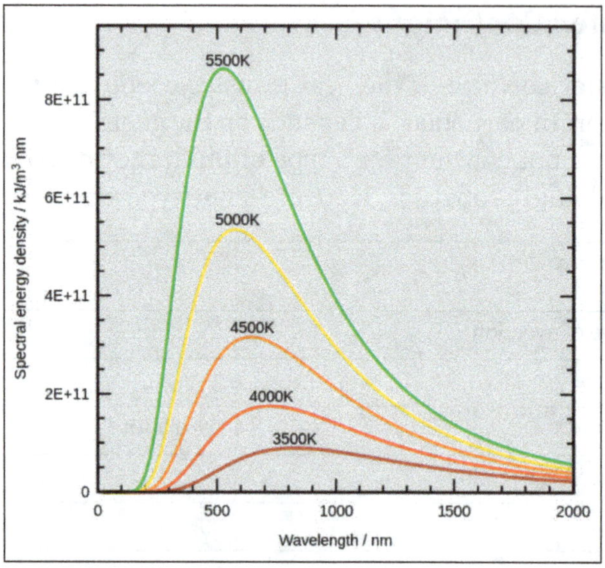

Relevant phenomena associated with photon interactions with surfaces are reflection, absorption, and transmission.

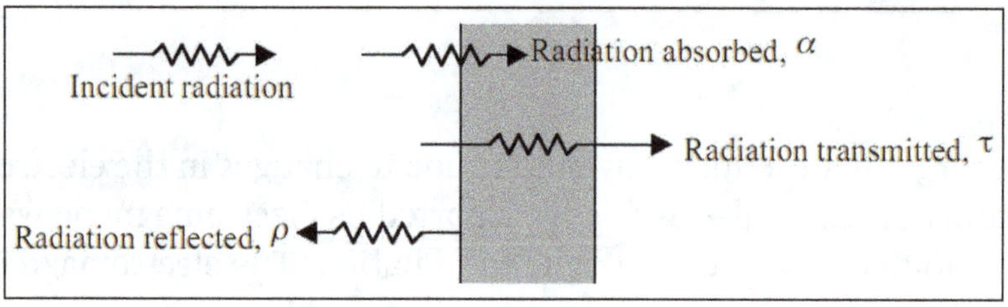

Describing Radiation Heat Transfer Quantitatively

Radiation rate equation. The net radiation heat flux from the surface due to exchange with the surroundings is,

$$\dot{Q}_{rad} = \epsilon \sigma A\left(T_s^4 - T_{sur}^4\right)$$

where,

$\dot{Q}_{rad}$	Convection heat-transfer rate through area A in the x direction between locations 1 and 2.	J/s W
A	Cross-sectional area through which conduction occurs.	m²
ϵ	The emmisivity of the surface.	Dimensionless
σ	The Stefan-Boltzmann constant.	5.67×10^{-8} W/m²K⁴
T_s	Temperature of the surface.	K
T_{sur}	Temperature of the surrounding surfaces.	K

T_∞ and T_{sur} are not the same. T_∞ indicates a free stream (liquid/gas) temperature that drives convection, while T_{sur} is the temperature of the surrounding surface involved in radiation.

Combined Mechanisms or Modes

Heat transfer mechanisms (conduction, convection, radiation) can often occur simultaneously. They can be occur in series and/or parallel. As we saw for mass transfer, resistance concepts can be useful when analyzing combined mechanisms, especially when they occur in series.

Resistances

The resistance is the driving potential (or driving force) divided by the transfer rate:

$$\text{resistance} = \frac{\text{driving potential}}{\text{transfer rate}}$$

The Effect of the Coordinate System

Although the driving force may remain the same, the expression for the transfer rate may vary based on the coordinate systems.

Cartesian Coordinate System

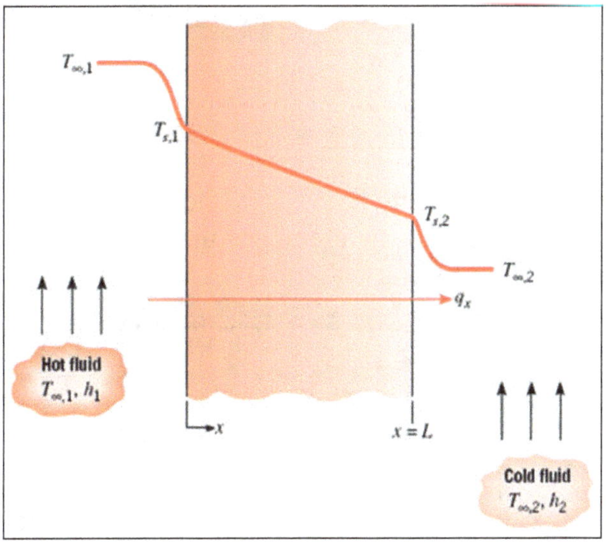

The heat transfer rate is,

$$Q_{cond} = \frac{kA}{L}\left(T_{S,1} - T_{S,2}\right)$$

Consequently, the heat transfer resistance due to conduction is,

$$R_{cond} = \frac{L}{kA}$$

The total resistance for this case (convection + conduction + convection) is,

$$R_{tot} = \frac{1}{h_1 A} + \frac{L}{kA} + \frac{1}{h_2 A}$$

Cylindrical Coordinate System

In the following, L is the length of the cylinder, not the appropriate distance over which radial conduction occurs, which is $r_2 - r_1$.

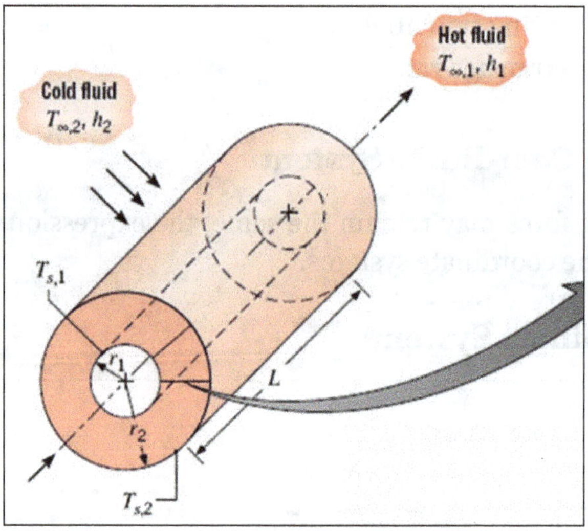

At steady state, the heat transfer rate across a cylinder of one radius must be equal to that for any other radius. This leads to a temperature profile in r that is not linear, but has a shape like this:

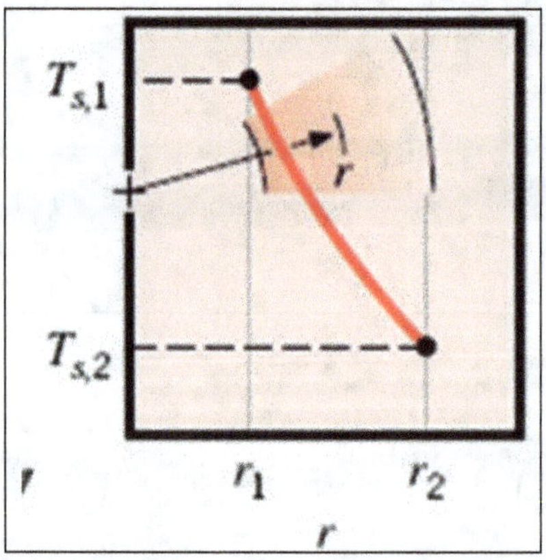

The equation for the temperature profile is,

$$T(r) = \frac{T_{s,1} - T_{s,2}}{\ln(r_1 / r_2)} \ln\left(\frac{r}{r_2}\right) + T_{s,2}$$

The corresponding heat transfer rate is,

$$\dot{Q} = \frac{2\pi Lk(T_{s,1} - T_{s,2})}{\ln(r_2 / r_1)}$$

and the heat transfer resistance is,

$$R_{cond} = \frac{\ln(r_2 / r_1)}{2\pi Lk}$$

The total resistance for this case (convection + conduction + convection) is,

$$R_{tot} = \frac{1}{h_1 A_1} + \frac{\ln(r_2 / r_1)}{2\pi Lk} + \frac{1}{h_2 A_2}$$

Stoichiometry

Material Balances on Reactive Processes

Accumulation = In – Out + Generation – Consumption

For a reactive process at steady-state, the general balance equation becomes,

$$0 = \text{In} - \text{Out} + \underbrace{\text{Generation} - \text{Consumption}}_{\text{Accounts for material produced or consumed within the system}}$$

The stoichiometric equation of the reaction imposes constraints on the relative amounts of reactants and products in the input and output streams. What quantities are conserved?

Type of Balance	Without Rxn	With Rxn
Total mass	✓	✓
Total moles	✓	✗
Mass of a chemical compound	✓	✗
Moles of a chemical compound	✓	✗
Mass of an atomic species	✓	✓
Moles of an atomic species	✓	✓

Basics of Stoichiometry

Stoichiometry is the theory of the proportions in which chemical species combine with one another in a reaction,

$$2\,SO_2 + O_2 \rightarrow 2\,SO_3$$

Stoichiometric equation is an equation that relates the relative number of molecules or moles of reactants and products (but not mass!) that participate in a chemical reaction. To be valid, the equation must be balanced. For example, are the following stoichiometric equations balanced?

$$C_2H_5OH + O_2 \rightarrow CO_2 + H_2O$$

$$(NH_4)_2\,Cr_2O_7 \rightarrow Cr_2O_3 + N_2 + H_2O$$

Stoichiometry Imposes Additional Constraints to Balances

- Defines relationship between generation of products and consumption of reactants.
- Always in terms of molar units (not mass).

Stoichiometric Coefficients (ν_i) are the values preceding each molecular species (i) in a balanced stoichiometric equation. Values are defined to be positive for products and negative for reactants. For the reaction:

$$2\,SO_2 + O_2 \rightarrow 2\,SO_3$$

$$\nu_{SO_2} = -2;\ \nu_{O_2} = -1;\ \nu_{SO_3} = 2$$

Stoichiometric Ratio is the ratio of stoichiometric coefficients in a balanced stoichiometric equation. For the above reaction for example,

$$\frac{2\ \text{mol}\ SO_3\ \text{generated}}{1\ \text{mol}\ O_2\ \text{consumed}}$$

Two reactants, A and B, are in stoichiometric proportion if the ratio (moles of A present)/(moles of B present) equals their stoichiometric ratio determined from the balanced stoichiometric equation.

Limiting and Excess Reactants

If the reactants are not present in stoichiometric proportions, the reactant that would be completely consumed if a reaction went to completion is known as the limiting reactant

(LR). The other reactant(s) are termed excess reactant(s). The fractional excess of the reactant is the ratio of the excess to the stoichiometric requirement:

$$\text{Fractional excess of A} = \frac{(n_A)_{feed} - (n_A)_{stoich}}{(n_A)_{stoich}} \text{ or } \left(= \frac{(\dot{n}_A)_{feed} - (\dot{n}_A)_{stoich}}{(\dot{n}_A)_{stoich}} \right)$$

where, $(n_A)_{feed}$ is the number of moles of an excess reactant, A, present in the feed to a reactor and $(n_A)_{stoich}$ is the stoichiometric requirement of A, or the amount needed to react completely with the limiting reactant, even if the reaction does not go to completion. Percentage excess is 100 times the fractional excess.

Fractional Conversion

Chemical reactions do not occur instantaneously, but often proceed quite slowly. Therefore, it is often not practical to design a reactor for complete conversion of the limiting reactant. Instead, the reactant is separated from the reactor outlet stream and recycled back to the reactor inlet. The fractional conversion of a reactant is the ratio of the amount reacted to the amount fed,

$$f_A = \frac{(n_A)_{reacted}}{(n_A)_{fed}} \quad \text{or} \quad f_A = \frac{(\dot{n}_A)_{reacted}}{(\dot{n}_A)_{fed}}$$

- If no reactant is specified, then fractional conversion refers to the limiting reactant (LR). This is usually the case.
- Fractional conversion is unitless. The percentage conversion is 100 × fA.
- The fraction unreacted (i.e., in exit stream) is 1 – fA.

"Fractional Conversion" can be considered as a process constraint (extra equation), as it describes how the process is operating (i.e. how much rxn did take place?).

Identifying the Limiting Reactant

- Balance the stoichiometric equation.
- Compare the ratios:

$$\left(\frac{n_x}{n_y}\right)_{feed} \text{ versus } \left(\frac{n_x}{n_y}\right)_{stoich}.$$

If, $\left(\frac{n_x}{n_y}\right)_{feed} \geq \left(\frac{n_x}{n_y}\right)_{stoich}$, reactant x is not the limiting reactant

If, $\left(\frac{n_x}{n_y}\right)_{feed} < \left(\frac{n_x}{n_y}\right)_{stoich}$, reactant x is the limiting reactant

How to Handle Different Stoichiometric Ratios?

The extent of reaction $(\xi \text{ or } \dot{\xi})$ is a quantity that characterizes the reaction and can significantly simplify calculations. For a continuous process at steady-state: $\dot{n}_i = \dot{n}_{io} + v_i\dot{\xi}$ where, $\dot{n}_{io}$ and $\dot{n}_i$ are the molar flow rates of species i in the feed and outlet streams, respectively (out = in +/- conversion). For a batch process: $n_i = n_{io} + v_i\xi$ How do these equations arise from the general M.B.? Where, n_{io} and n_i are the initial and final molar amounts of species i, respectively. The extent of reaction $(\xi \text{ or } \dot{\xi})$ has the same units as n (or $\dot{n}$). Every reaction has a single, unique value of ξ.

Ideal Gas Law

Before we look at the Ideal Gas Equation, let us state the four gas variables and one constant for a better understanding. The four gas variables are: pressure (P), volume (V), number of mole of gas (n), and temperature (T). Lastly, the constant in the equation shown below is R, known as the the gas constant:

$$PV = nRT$$

Another way to describe an ideal gas is to describe it in mathematically. Consider the following equation:

$$\frac{PV}{nRT} = 1$$

The term $\frac{PV}{nRT}$ is also called the compression factor and is a measure of the ideality of the gas. An ideal gas will always equal 1 when plugged into this equation. The greater it deviates from the number 1, the more it will behave like a real gas rather than an ideal. A few things should always be kept in mind when working with this equation.

- Pressure is directly proportional to number of molecule and temperature. (Since P is on the opposite side of the equation to n and T).
- Pressure, however, is indirectly proportional to volume. (Since P is on the same side of the equation with V).

Simple Gas Laws

The Ideal Gas Law is simply the combination of all Simple Gas Laws (Boyle's Law, Charles' Law, and Avogadro's Law), and so learning this one means that you have learned them all. The Simple Gas Laws can always be derived from the Ideal Gas equation.

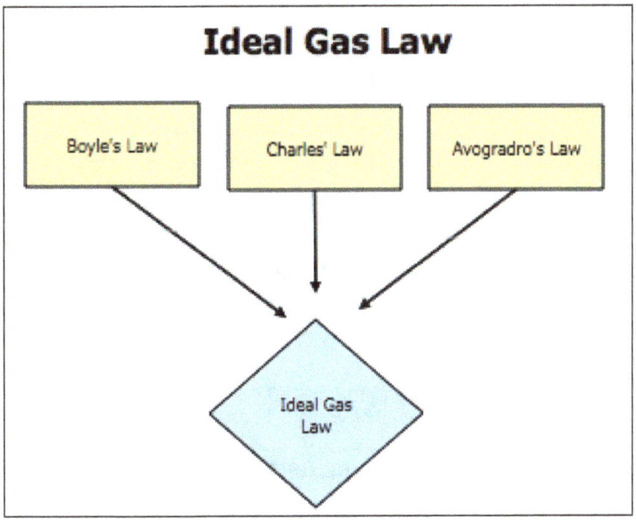

Boyle's Law

Boyle's Law describes the inverse proportional relationship between pressure and volume at a constant temperature and a fixed amount of gas. This law came from a manipulation of the Ideal Gas Law.

$$P \propto \frac{1}{V}$$

Or expressed from two pressure/volume points:

$$P_1V_1 = P_2V_2$$

This equation would be ideal when working with problem asking for the initial or final value of pressure or volume of a certain gas when one of the two factor is missing.

Charles's Law

Charles's Law describes the directly proportional relationship between the volume and temperature (in Kelvin) of a fixed amount of gas, when the pressure is held constant.

$$V \propto T$$

Or express from two volume/temperature points:

$$\frac{V_1}{T_1} = \frac{V_2}{T_2}$$

This equation can be used to solve for initial or final value of volume or temperature under the given condition that pressure and the number of mole of the gas stay the same.

Avogadro's Law

Volume of a gas is directly proportional to the amount of gas at a constant temperature and pressure.

$$V \propto n$$

Or expressed as a two volume/number points:

$$\frac{V_1}{n_1} = \frac{V_2}{n_2}$$

Avogadro's Law can apply well to problems using Standard Temperature and Pressure, because of a set amount of pressure and temperature.

Amontons's Law

Given a constant number of mole of a gas and an unchanged volume, pressure is directly proportional to temperature.

$$P \propto T$$

Or expressed as two pressure/temperature points:

$$\frac{P_1}{T_1} = \frac{P_2}{T_2}$$

Boyle's Law, Charles' Law, and Avogradro's Law and Amontons's Law are given under certain conditions so directly combining them will not work. Through advanced mathematics (provided in outside link if you are interested), the properties of the three simple gas laws will give you the Ideal Gas Equation.

Standard Temperature and Pressure (STP)

Standard condition of temperature and pressure is known as STP. Two things you should know about this is listed below:

- The universal value of STP is 1 atm (pressure) and 0 °C. Note that this form specifically stated 0 °C degree, not 273 Kelvin, even though you will have to convert into Kelvin when plugging this value into the Ideal Gas equation or any of the simple gas equations.
- In STP, 1 mole of gas will take up 22.4 L of the volume of the container.

Units of P, V and T

The table below lists the different units for each property.

Factor	Variable	Units
Pressure	P	atm Torr Pa mmHg
Volume	V	L m^3
Moles	n	mol
Temperature	T	K
Gas Constant	R*	see Values of R table below

Take note of certain things such as temperature is always in its SI units of Kelvin (K) rather than Celsius (C), and the amount of gas is always measured in moles. Gas pressure and volume, on the other hand, may have various different units, so be sure to know how to convert to the appropriate units if necessary.

Pressure Units

Use the following table as a reference for pressure.

Common Units of Pressure		
Unit	Symbol	Equivalent to 1 atm
Atmosphere	atm	1 atm
Millimeter of Mercury	mmHg	760 mmHg
Torr	Torr	760 Torr
Pascal	Pa	101326 Pa
Kilopascal	kPa*	101.326 kPa
Bar	bar	1.01325 bar
Millibar	mb	1013.25 mb

*This is the SI unit for pressure.

The Gas Constant (R)

Value of R WILL change when dealing with different unit of pressure and volume (Temperature factor is overlooked because temperature will always be in Kelvin instead of Celsius when using the Ideal Gas equation). Only through appropriate value of R will you get the correct answer of the problem. It is simply a constant and the different values of R correlates accordingly with the units given. When choosing a value of R, choose the one with the appropriate units of the given information (sometimes given units must be converted accordingly). Here are some commonly used values of R:

Values of R
0.082057 L atm mol^{-1} K^{-1}
62.364 L Torr mol^{-1} K^{-1}
8.3145 m^3 Pa mol^{-1} K^{-1}
8.3145 J mol^{-1} K^{-1*}

Process Variables

Here is a list of a few of the symbols and definitions:

Symbol	Definition	Units
m	Mass	Mass
m_A	Mass of species a	Mass
n	Moles	Moles
n_A	Moles of species a	Moles
V	Volume of a process 'container'	Volume
MW_A	Molecular weight of species a	Mass/mole
$\dot{m}$	Mass flow rate	Mass/time
$\dot{n}$	Molar flow rate	Moles/time
$\dot{V}$	Volumetric flow rate	Volume/time
x_A	Mass fraction of species a	Unitless (mass/mass)
y_A	Mole fraction of species a	Unitless (moles/moles)
c_A	Concentration of species a	Moles/volume
ρ	Density	Mass/volume
P	Pressure	Force/area
T	Temperature	Temperature

Note that overdots represent rates (quantity per unit time). More specifically, for the rates, we have the following definitions:

- Mass flow rate $\dot{m}$: The mass of a material that passes a reference plane within a unit time interval.
- Molar flow rate $\dot{n}$: The number of moles of a material that passes a reference plane within a unit time interval.

- Volumetric flow rate $\dot{V}$: The volume of a material that passes a reference plane within a unit time interval.

Useful Relationships between Some of these Variables

These relationships will help convert 'what you know' into 'what you need' to formulate the equations and solve for the quantities of interest.

$$m = \rho V$$

$$x_A = \frac{m_A}{m}$$

$$y_A = \frac{n_A}{n}$$

$$\dot{m} = \rho \dot{V}$$

$$c_A = \frac{n_A}{V}$$

$$\dot{m}_A = x_A \dot{m} = MW_A \dot{n}_A = MW_A y_A \dot{n} = MW_A c_A \dot{V}$$

$$\dot{n}_A = \frac{\dot{m}_A}{MW_A} = \frac{x_A \dot{m}}{MW_A} = y_A \dot{n} = c_A \dot{V}$$

Converting between Mole Fraction and Mass Fraction

In general, it is useful to assume some amount of materials for the calculation. We call this the basis of calculation or just basis.

Mole Fraction to Mass Fraction

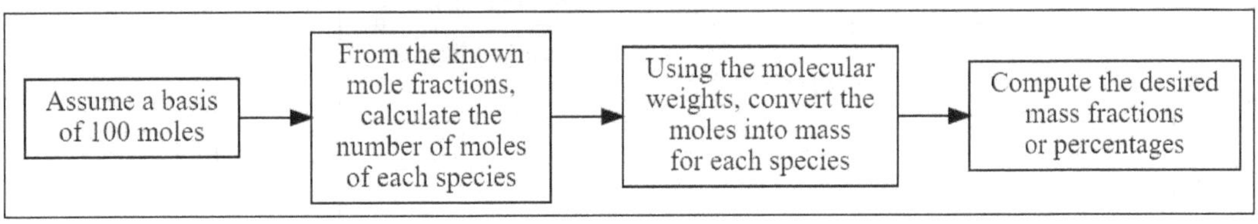

Mass Fraction to Mole Fraction

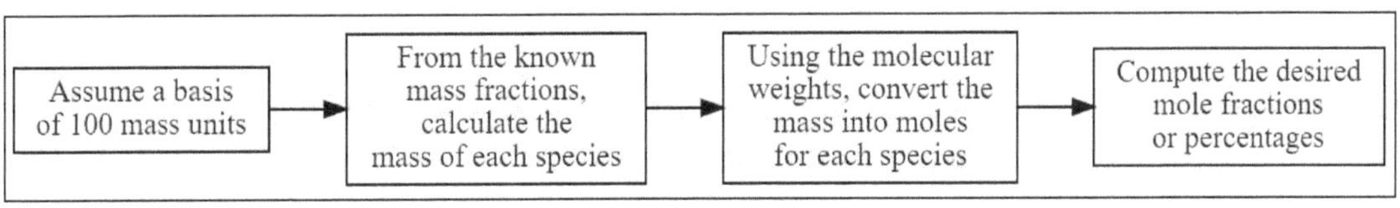

Reality Check on Results

The mass and mole fractions must sum to 1. So, if N is the number of chemical species,

$$\sum_{i}^{N} x_i = 1$$

$$\sum_{i}^{N} y_i = 1$$

The Stream Table

The stream table lists important variables for all process streams in an organized, tabular format. The rows and columns are customized for the application. Typically, rows would be variables like composition, flow rate, temperature, and pressure; and columns would normally correspond to points before and after pieces of equipment. A process stream table is also called a process flow-sheet.

Heat and Material Balance Table										
Stream ID		AIR	FEED	NAT-GAS	PRIM-IN	PRIM-OUT	PROD	SEC-IN	SEC-OUT	STEAM
Temperature	C	600,0	32,7	33,0	600,0	800,0	173,0	764,3	995,8	33,0
Pressure	BAR	35,290	39,220	39,220	39,220	35,290	35,290	35,290	35,290	39,220
Vapour Frac		1,000	0,251	1,000	1,000	1,000	0,913	1,000	1,000	0,000
Mole Flow	KMoL/HR	1729,330	5004,443	1254,443	500,4443	6526,358	8828,312	8255,688	8828,312	3750,000
Mass Flow	KG/HR	50097,051	87783,203	20225,902	87783,203	87783,203	137880,250	137880,250	137880,250	67557,297
Volume Flow	CUM/HR	3594,061	822,613	754,243	9150,887	16582,898	8414,034	20298,414	26545,707	68,533
Enthalpy	MMKCAL/HR	7,250	-279,729	-22,649	-212,141	-162,151	-222,490	-154,901	-154,902	-257,080
Mole Flow	KMoL/HR									
CH_4			1250,000	1250,000	1250,000	489,042	21,281	489,042	21,281	
H_2O			3750,000		3750,000	2600,897	2832,839	2600,897	2832,839	3750,000
H_2						2671,019	3374,600	2671,019	3371,600	
Co						372,812	814,477	372,812	814,477	
Co_2			3,000	3,000	3,000	391,145	417,242	391,145	417,242	
N_2		1349,150	1,443	1,443	1,443	1,443	1350,593	1350,593	1350,593	
O_2		362,900						362,900		
AR		17,280					17,280	17,280	17,280	

Flow Rates

Almost every process involves the movement of material from one point to another. The movement of material is present between process units, production facilities, and transportation depots. The rate at which materials are transported is the flow rate.

Mass Flow Rate

Mass flow rate is the measurement of mass per unit time (mass/time). This flowrate is usually denoted by the character $\dot{m}$, where $\dot{m} = \frac{kg}{s}$. Since mass must be conserved, the input masses of a process must equal the output masses, unless there is accumulation. Therefore, mass flowrate is a very useful tool in solving material balances.

Molar Flow Rate

Molar flow rate is the measurement of moles per unit time (mole/time). Since the total number of moles of atomic species (eg. N, O, C) must be conserved it is useful in solving both reactive and non-reactive problems.

Volumetric Flow Rate

Volumetric flow rate is the measurement of volume per unit time (volume/time) This flow rate is usually denoted by the character $\dot{V}$, where $\dot{V} = \frac{m^3}{s}$. Volumetric flow rate may change with density.

$$\rho = \frac{m}{V} = \frac{\dot{m}}{\dot{V}} = \frac{\frac{kg}{s}}{\frac{m^3}{s}} = \frac{kg}{m^3}$$

$$\dot{m} = \rho \dot{V}$$

Since mass flow rate is more useful, but volumetric flow rate is easier to measure, we use the formula shown above to convert from one basis to another. Although density is nearly conserved for solids and liquids, it is not for gases. For example, if a gas is flowing through a nozzle, the mass flow rate will not change but the gas will compress or expand.

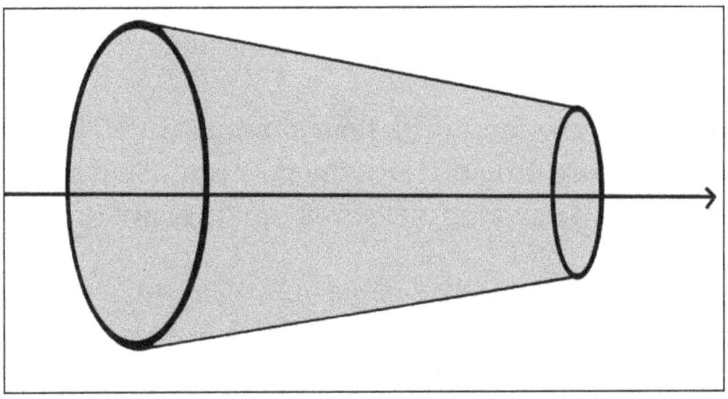

Flow Rate Measurement

Flow rates are measured by a device called a flowmeter. This device provides a continuous reading of the flow rate in the line. Two commonly used flowmeters are rotameters and orifice meters.

Rotameters

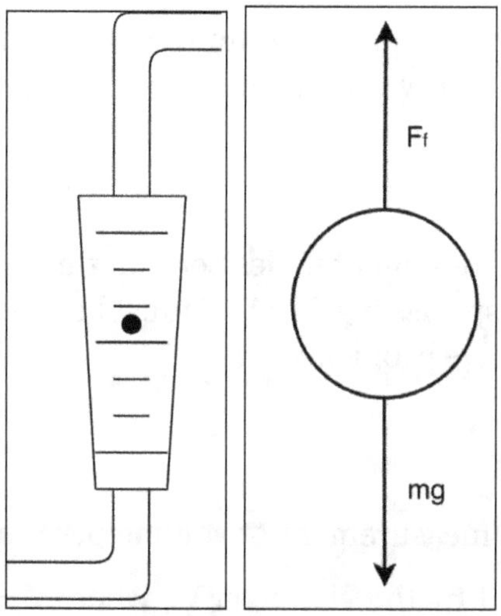

Rotameters are a tapered vertical tube containing a small ball (the black dot in the image). The larger the flow rate, the higher the ball rises due to the increased friction force.

Orifice Meter

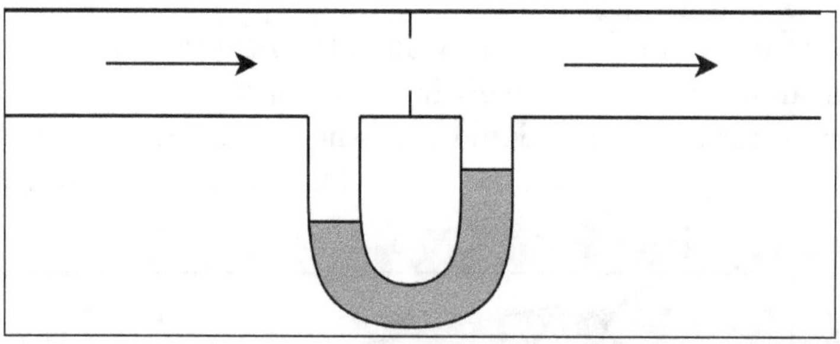

The orifice meter is an obstruction in the flow channel which causes a pressure drop. The magnitude of the pressure drop dictates the flow rate. The larger the pressure drop, the larger the flow rate. A manometer measures the pressure drop.

Manometer

Manometers are one of the earliest tools used in measuring pressure drops. This is done by measuring the heights of the different fluids in the manometer.

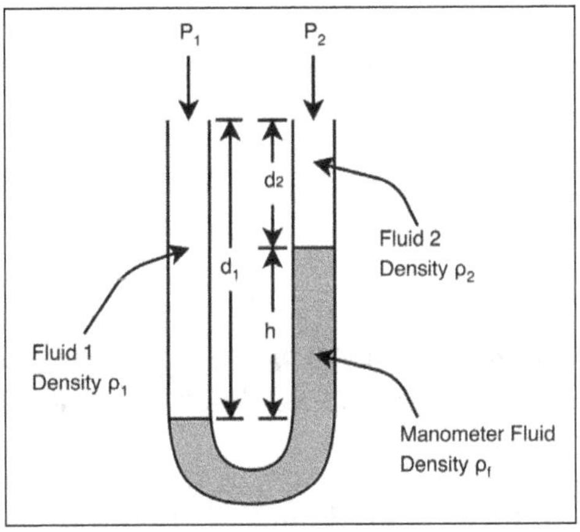

Writing and equation the expressions for the pressures at points (a) and (b) in the figure shown above, we get the general manometer equation:

$$P_1 + \rho_1 g d_1 = P_2 + \rho_2 g d_2 + \rho_f g h \left(\text{General Manometer Equation}\right)$$

If the densities of fluids 1 and 2 are the same, the equation simplifies to:

$$P_1 - P_2 = (\rho_f - \rho) g h \left(\text{Differential Manometer Equation}\right)$$

If fluids 1 and 2 are gases, the equation simplifies further to:

$$P_1 - P_2 = \rho_f g h \left(\text{Manometer Formula for Gases}\right).$$

Degrees of Freedom

A degree of freedom of a physical system refers to a (typically real) parameter that is necessary to characterize the state of a physical system. Consider a point particle that is free to move in three dimensions. The location of any particle in three dimensional space can be specified by three position coordinates: x, y, and z. The direction and speed at which a particle moves can be described in terms of three velocity components, e.g. vx, vy, and vz. If the time evolution of the system is deterministic, where the state at one instant uniquely determines its past and future position and velocity as a function of time, such a system will have six degrees of freedom. If the motion of the particle is constrained to a lower number of dimensions – if, for example, the particle must move along a wire or on a fixed surface – then the system will have less than six degrees of freedom. On the other hand, a system with an extended object that may rotate or vibrate can have more than six degrees of freedom. A force on the particle that depends only upon time and the particle's position and velocity fits this description.

In mechanics, a point particle's state at any given time can be described with position and velocity coordinates in the Lagrangian formalism, or with position and momentum coordinates in the Hamiltonian formalism.

Similarly, in statistical mechanics, a degree of freedom is a single scalar number describing the microstate of a system. The specification of all microstates of a system is a point in the system's phase space.

A degree of freedom may be any useful property that is not dependent on other variables. For example, in the 3D ideal chain model, two angles are necessary to describe each monomer's orientation.

In statistical mechanics and thermodynamics, it is often useful to specify quadratic degrees of freedom. These are degrees of freedom that contribute in a quadratic way to the energy of the system. They are also variables that contribute quadratically to the Hamiltonian.

Degrees of Freedom of Gas Molecules

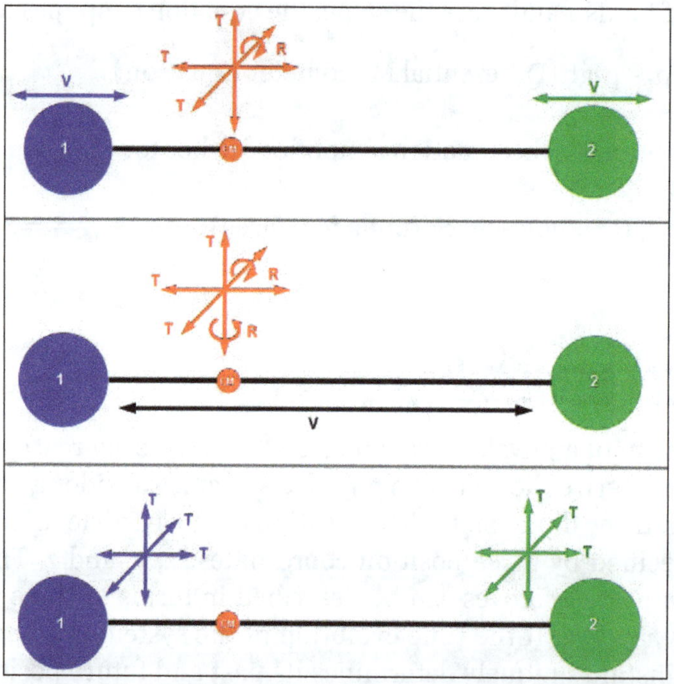

Different ways of visualizing the 6 degrees of freedom of a diatomic molecule. (CM: center of mass of the system, T: translational motion, R: rotational motion, V: vibrational motion.)

In three-dimensional space, three degrees of freedom are associated with the movement of a particle. A diatomic gas molecule thus has 6 degrees of freedom. This set may be decomposed in terms of translations, rotations, and vibrations of the molecule. The center of mass motion of the entire molecule accounts for 3 degrees of freedom. In addition, the molecule has two rotational degrees of motion and one vibrational mode.

The rotations occur around the two axes perpendicular to the line between the two atoms. The rotation around the atom–atom bond is not a physical rotation. This yields, for a diatomic molecule, a decomposition of,

$$3N = 6 = 3 + 2 + 1.$$

For a general (non-linear) molecule with N > 2 atoms, all 3 rotational degrees of freedom are considered, resulting in the decomposition,

$$3N = 3 + 3 + (3N - 6)$$

Which means that an N-atom molecule has 3N – 6 vibrational degrees of freedom for N > 2. In special cases, such as adsorbed large molecules, the rotational degrees of freedom can be limited to only one. One can also count degrees of freedom using the minimum number of coordinates required to specify a position. This is done as follows:

- For a single particle we need 2 coordinates in a 2-D plane to specify its position and 3 coordinates in 3-D plane. Thus its degree of freedom in a 3-D plane is 3.
- For a body consisting of 2 particles (ex. a diatomic molecule) in a 3-D plane with constant distance between them (let's say d) we can show (below) its degrees of freedom to be 5.

Let's say one particle in this body has coordinate (x_1, y_1, z_1) and the other has coordinate (x_2, y_2, z_2) with z_2 unknown. Application of the formula for distance between two coordinates $d = \sqrt{(x_2 - x_1)^2 + (y_2 - y_1)^2 + (z_2 - z_1)^2}$ results in one equation with one unknown, in which we can solve for z_2 one of x_1, x_2, y_1, y_2, z_1 or z_2 can be unknown.

Contrary to the classical equipartition theorem, at room temperature, the vibrational motion of molecules typically makes negligible contributions to the heat capacity. This is because these degrees of freedom are frozen because the spacing between the energy eigenvalues exceeds the energy corresponding to ambient temperatures (kBT). Such degrees of freedom are disregarded because of their low effect on total energy. However, at very high temperatures they cannot be neglected.

Independent Degrees of Freedom

The set of degrees of freedom X_1, ... , XN of a system is independent if the energy associated with the set can be written in the following form:

$$E = \sum_{i=1}^{N} E_i(X_i),$$

where, E_i is a function of the sole variable X_i.

Example: if X_1 and X_2 are two degrees of freedom, and E is the associated energy:

- If $E = X_1^4 + X_2^4$, then the two degrees of freedom are independent.
- If $E = X_1^4 + X_1X_2 + X_2^4$ then the two degrees of freedom are not independent. The term involving the product of X_1 and X_2 is a coupling term that describes an interaction between the two degrees of freedom.

For i from 1 to N, the value of the ith degree of freedom X_i is distributed according to the Boltzmann distribution. Its probability density function is the following:

$$p_i(X_i) = \frac{e^{-\frac{E_i}{k_B T}}}{\int dX_i e^{-\frac{E_i}{k_B T}}}$$

The brackets $\langle\rangle$ denote the mean of the quantity they enclose. The internal energy of the system is the sum of the average energies associated to each of the degrees of freedom:

$$\langle E \rangle = \sum_{i=1}^{N} \langle E_i \rangle$$

Quadratic Degrees of Freedom

A degree of freedom X_i is quadratic if the energy terms associated to this degree of freedom can be written as,

$$E = \alpha_i X_i^2 + \beta_i X_i Y$$

where Y is a linear combination of other quadratic degrees of freedom.

Example: if X_1 and X_2 are two degrees of freedom, and E is the associated energy:

- If $E = X_1^4 + X_1^3 X_2 + X_2^4$, then the two degrees of freedom are not independent and non-quadratic.
- If $E = X_1^4 + X_2^4$, then the two degrees of freedom are independent and non-quadratic.
- If $E = X_1^2 + X_1X_2 + 2X_2^2$, then the two degrees of freedom are not independent but are quadratic.
- If $E = X_1^2 + 2X_2^2$, then the two degrees of freedom are independent and quadratic.

For example, in Newtonian mechanics, the dynamics of a system of quadratic degrees of freedom are controlled by a set of homogeneous linear differential equations with constant coefficients.

Quadratic and Independent Degree of Freedom

X_1, ..., XN are quadratic and independent degrees of freedom if the energy associated to a microstate of the system they represent can be written as:

$$E=\sum_{i=1}^{N}\alpha_i X_i^2$$

Equipartition Theorem

In the classical limit of statistical mechanics, at thermodynamic equilibrium, the internal energy of a system of N quadratic and independent degrees of freedom is:

$$U=\langle E\rangle=N\frac{k_B T}{2}$$

Here, the mean energy associated with a degree of freedom is:

$$\langle E_i\rangle=\int dX_i\,\alpha_i X_i^2\,p_i(X_i)=\frac{\int dX_i\,\alpha_i X_i^2\,e^{-\frac{\alpha_i X^2}{k_B T}}}{\int dX_i\,e^{-\frac{\alpha_i X^2}{k_B T}}}$$

$$\langle E_i\rangle=\frac{k_B T}{2}=\frac{\int dx\,x^2\,e^{-\frac{x^2}{2}}}{\int dXe^{-\frac{x^2}{2}}}=\frac{k_B T}{2}$$

Since the degrees of freedom are independent, the internal energy of the system is equal to the sum of the mean energy associated with each degree of freedom, which demonstrates the result.

Generalizations

The description of a system's state as a point in its phase space, although mathematically convenient, is thought to be fundamentally inaccurate. In quantum mechanics, the motion degrees of freedom are superseded with the concept of wave function, and operators which correspond to other degrees of freedom have discrete spectra. For example, intrinsic angular momentum operator (which corresponds to the rotational freedom) for an electron or photon have only two eigenvalues. This discreteness becomes apparent when action has an order of magnitude of the Planck constant, and individual degrees of freedom can be distinguished.

References

- What-is-chemical-engineering, about-the-program, undergraduate-students, prospective-students: utexas.edu, Retrieved 29, June 2020
- Heat-transfer: engr.colostate.edu, Retrieved 09, January 2020
- The-Ideal-Gas-Law, Properties-of-Gases, Physical-Properties-of-Matter, States-of-Matter Supplemental-Modules-(Physical-and-Theoretical-Chemistry): libretexts.org, Retrieved 12, August 2020
- Process-variables: engr.colostate.edu, Retrieved 05, May 2020
- Module-1-Proces-Basics-Water: github.io, Retrieved 17, February 2020

2

Chemical Processes and Reactions

There are various principles and concepts which are central to the discipline of chemical engineering. Some of these are membrane process, adsorption, crystallization process, nitration, and sulfonation and sulfation. These diverse principles and concepts of chemical engineering have been thoroughly discussed in this chapter.

Processes are the main concern of chemical engineers. A process is a system that converts feed stocks of lower intrinsic value to products of higher value. For instance, the block diagram of a process to manufacture ammonia from natural gas is shown in Figure. This process has several sections, each one of which carries out a specific task and is, in effect, a mini-process. Natural gas, steam and air are fed to the Reformer Section that converts these feeds into a mixture of H_2, CO,

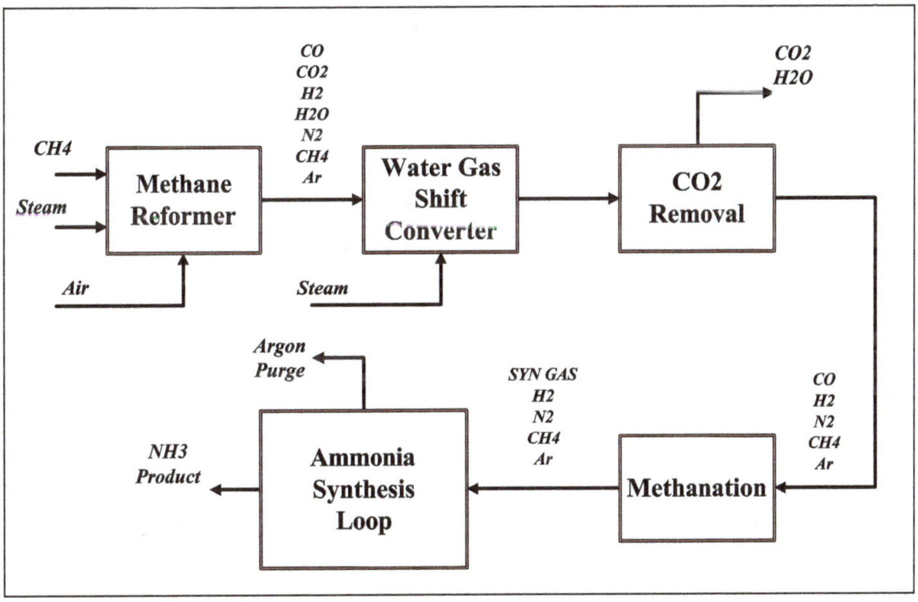

Block Diagram for Ammonia Process.

CO_2, N_2, and H_2O. The overall chemical reactions involved are:

$$CH_4 + H_2O \rightarrow CO + 3\,H_2$$

$$CH_4 + O_2 \rightarrow CO + 2\,H_2$$

$$CH_4 + 2\,O_2 \rightarrow CO_2 + 2\,H_2O$$

Since H_2 is the desired raw material from which to make ammonia, this gas mixture is sent to the CO Shift Section where additional steam is added to improve conversion by the water gas shift reaction:

$$CO + H_2O \rightarrow H_2 + CO_2$$

The CO_2 and H_2O present in the gas mixture leaving the CO Shift Section are removed in the CO_2 Removal Section. The gas mixture leaving the CO_2 Removal Section contains primarily a 3/1 mixture of H_2 and N_2 (the N_2 coming from the air fed to the Reformer Section). It also contains small amounts of CO (the water gas shift reaction is equilibrium limited) as well as argon from the air feed to the Reformer Section.

The CO must be removed from this mixture because it will deactivate the catalyst used in the ammonia converter. This is done in the Methanation Section via the reaction.

$$CO + 3\,H_2 \rightarrow CH_4 + H_2O$$

The gas mixture leaving the Methanation Section contains a 3/1 mixture of H_2 and N_2 and trace amounts of CH_4 and Ar. This is sent to the NH_3 Synthesis Loop where the ammonia is made via the well-known reaction,

$$N_2 + 3\,H_2 \rightarrow 2\,NH_3.$$

So far each section of the process has been considered as a black box. Let us look at the Ammonia Synthesis Loop in more detail. A process flow diagram (PFD) is shown in Figure. The ammonia synthesis reaction is equilibrium limited to the point where it must be run at rather high pressures (3000 PSIA or 20 mPa) in order to achieve a reasonable conversion of Syn Gas to ammonia across the Converter. Since the front end of the process is best run at much lower pressures, a Feed Compressor is needed to compress the Syn Gas to the operating pressure of the Synthesis Loop.

Due to the unfavorable reaction equilibrium, only part of the Syn Gas is converted to ammonia on a single pass through the Converter. Since the unconverted Syn Gas is valuable, the majority of it is recycled back to the Converter. Due to pressure drop through the Synthesis Loop equipment, a Recycle Compressor is required to make up this pressure drop. The recycled Syn Gas is mixed with fresh Syn Gas to provide the feed to the Synthesis Converter.

The effluent from the Synthesis Converter contains product ammonia as well as unreacted Syn Gas. These must be separated. At the pressure of the synthesis loop ammonia can be condensed at reasonable temperatures. This is done in the Ammonia Condenser where the Syn Gas is cooled to approximately ambient temperature using cooling water (CW). The liquid ammonia is then. Separated from the cycle gas in the Ammonia Knockout (KO) Drum. The liquid ammonia is removed from the bottom of the drum and the cycle gas leaves the top.

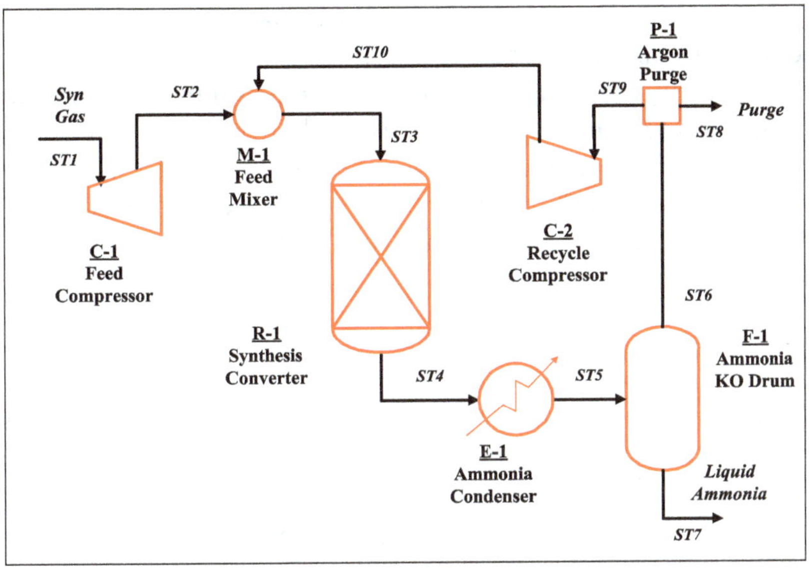

Ammonia Synthesis Loop Process Flow Diagram.

Some of the cycle gas must be purged from the Synthesis Loop. Otherwise, the argon that enters the loop in the Syn Gas has no way to leave and will build up in concentration. This will reduce the rate of the ammonia synthesis reaction to an unacceptable level. To prevent this from happening, a small amount of the cycle gas must be purged, the amount being determined by the amount of argon in the feed and its acceptable level in the Synthesis Converter feed (generally about 10 mol %).

This description of the Ammonia Synthesis Loop covers only the most important aspects. Modern ammonia plants are much more complex due to attention paid to maximizing the amount of ammonia produced per mol of Syn Gas fed to the loop. However, the process shown in Figure II-2 is typical of many chemical processes.

Basic Processing Functions

Several basic processing activities are required by the Ammonia Synthesis Loop in order to convert the hydrogen and nitrogen to ammonia product. These activities are common to almost all chemical processes; the functionality of each can be considered independently of any specific process. There are five of these processing activities that are of major interest in chemical engineering:

- Chemical Reaction,
- Mixing,
- Separation,
- Materials Transfer (Fluid flow),
- Energy (Heat) Transfer.

Of these the first three are involved in the process material balance. The last two are necessary adjuncts to operation of chemical processes. Material must be transported from one piece of equipment to another. For the large number of chemical plants processing only liquids and vapors, this involves fluid flow. Also, streams must be heated to or cooled to specified temperatures as dictated by the needs of the process. For example, reactors in general are operated at temperatures higher than those that are acceptable for most separation operations. Thus, streams must be heated to reaction temperature and then cooled back down for subsequent processing.

Chemical Reaction

What distinguishes the chemical process industries from almost all others is the use of chemical reactions to convert less valuable raw materials to more valuable products. In other words, chemical reaction is the heart and soul of almost all processes. In the Ammonia Synthesis Loop example, R-1, the Synthesis Converter is where the ammonia synthesis reaction takes place.

Mixing

Many chemical reactions involve two or more reactants. In order for these reactants to react, these must be brought into contact at the molecular level, i.e., mixed, before the desired reactions can proceed properly.

Mixing is also required if several substances are to be blended to create a product mixture with the desired properties. For our example process, mixing of fresh synthesis gas and recycled synthesis gas takes place in the Feed Mixer, M-1, before being sent to the Synthesis Converter.

Separation

In an ideal chemical process, exactly the right amounts of reactants would be mixed and reacted completely to the desired product. Unfortunately, this is seldom the case. Many reactions cannot be carried to completion for various reasons. Seldom do the reactants react only to the desired product. Unwanted by products are formed in addition to the target product. Finally, the reactants are seldom 100% pure, again for many reasons.

The result is that the material leaving the reactor is a mixture containing the desired product, by-products, unreacted raw materials, and impurities. It may also contain other components deliberately introduced for one reason or another. The use of a homogeneous catalyst is just one example of this.

This mixture must separate into its various constituents. The product must be separated from almost everything else and brought to an acceptable level of purity such that it can be sold. The reactants, being valuable, must be recovered and recycled back to the reactor. The impurities and by-products must be separated out for disposal in a suitable manner.

Thus, the activity of separation is fundamental to the operation of almost any process. Some separation systems are relatively simple. Others constitute the major part of the process. As will be seen, separation takes on many forms.

Separation of ammonia from cycle gas takes place in the Ammonia KO Drum, F-1. The cycle gas leaving the Ammonia Condenser, E-1, contains droplets of liquid ammonia. This mixture enters the middle of the KO Drum. The vapor, being less dense, flows upward while the liquid ammonia falls to the bottom of the drum.

Materials Transfer

The Ammonia Synthesis Loop consists of several items of equipment, each of which has material flowing in and material flowing out. These flows take places though process piping connecting the various items of equipment. If the pressure in the upstream item of equipment is sufficiently higher than that in the downstream item, then material will flow from the upstream equipment to that downstream without the need for any additional equipment. This pressure difference is necessary to overcome the friction due to fluid flow through the piping. As a result, the pressure will decrease in the direction of flow through the process.

Thus, as synthesis gas flows from the inlet to the reactor (Stream 3) through the reactor, condenser and knockout drum, the pressure decreases significantly. In order to be able to recycle the unreacted hydrogen and nitrogen back to the reactor, some means is required to increase the pressure of this stream back to that of the reactor inlet. Hence the presences of the compressor in the recycle loop.

Unit Operations

One of the major contributions to the practice of chemical engineering is the concept of the unit operations. This concept was developed by Arthur D. Little and Warren K. Lewis in the early 1900's. Prior to its development, chemical engineering was, to a large extent, practiced along the lines of specific process technologies. For instance, if distillation was required in the manufacture of acetic acid, it became a problem in acetic acid distillation. The fact that a similar distillation might be required for the manufacture of, say, acetaldehyde was largely ignored. What Little and Lewis did was to show that the principles of distillation (as well as many other processing operations) were the same regardless of the materials being processed. So, if one knew how to design distillation columns, one could do so for acetic acid, acetaldehyde, or any other mixture of reasonable volatility with equal facility. The same proved to be true of other operations such as heat transfer by two-fluid heat exchangers, gas compression, liquid pumping, gas absorption, liquid-liquid extraction, fluid mixing, and many other operations common to the chemical industry.

The case for chemical reactors is less clear. Each reaction system tends to be somewhat unique in terms of its reaction conditions (temperature, pressure, type of catalyst, feed composition, residence time, heat effects, and equilibrium limitations). Thus, each reaction system must be approached on its own merits with regard to the choice of reactor type and design. However, the field of chemical reaction engineering has undergone substantial development over the past few decades, the result being that much of what is required for the choice and design of reactors is subject to a rational and quantitative approach.

Modes of Process Operation

As mentioned previously, there are several modes of process operation. The one that has been most widely studied by chemical engineers is that of the steady-state operation of continuous processes. The reasons have also been discussed. In this mode of operation we assume that the process is subject to such good control that as feed materials flow into the process as constant flow rates, the necessary reactions, separations, and other operations all take place at conditions that do not vary with time. The amount of material in each item of equipment (its inventory) does not vary with time; pressures and liquid levels are constant. Nor do temperatures, compositions, and flow rates at each point in the process vary with time. From the standpoint of the casual observer, nothing is happening. However, finished product is flowing out of the other end of the process into the product storage tanks.

The next most common mode of process operation is known as batch operation. Here, every processing operation is carried out in a discrete step. Reactants are pumped into a reactor, mixed, and heated up to reaction temperature. After a suitable length of time, the reactor turned off by cooling it down. It now contains a mixture of products, by-products, and unreacted reactants. These are pumped out of the reactor to the first of the various separation steps, possibly a batch distillation or a filtration if one of the products or by products is a solid. Each batch step has a beginning, time duration, and an end. (Most activities around the home are batch in nature - cooking, washing clothes, etc.)

A third mode is cyclic operation. From the standpoint of flows in and out of both the process and individual items of equipment, operation is continuous. However, in one or more items of equipment, operating conditions vary in time in a cyclical manner. A typical example is reactor whose catalyst deactivates fairly rapidly with time due to, say, coke formation on the catalyst. To recover the catalyst activity, it must be regenerated by being taken out of reaction operation. The coke is removed either by stripping by blowing an inert gas over the catalyst or, in the more difficult cases, by burning the coke off with dilute oxygen in an inert gas carrier.

Two characteristics of cyclical operation become apparent. First off, if the process is to be operated continuously but it one or more items of equipment must be taken off line for regeneration of one sort or another we must have at least two of such items available in parallel. One is on line while another is being regenerated. The second characteristic is that the operation conditions in the cyclically operated equipment must vary with time. If catalyst activity decreases with time, then something must be done to maintain the productivity of the reactor. Usually this is achieved by raising the reactor temperature. Thus, a freshly regenerated reactor will start off at a relatively low temperature; the temperature will be raised during the cycle; and the reactor will be taken off line when no further benefit is to be obtained by raising the temperature any further.

Membrane Process

Membrane processes cover a group of separation processes in which the characteristics of a membrane (porosity, selectivity, electric charge) are used to separate the components of a solution or a suspension. In these processes the feed stream is separated into two: the fraction that permeates through the membrane, called the permeate, and the fraction containing the components that have not been transported through the membrane, usually called the retentate The size of the components to be separated and the nature and magnitude of the driving force provide criteria for a classification of the membrane separation processes, as shown in the table. It should be noted that the boundaries between some of the processes, such as reverse osmosis and ultrafiltration, are arbitrary.

Of the processes listed in Table, reverse osmosis and ultafiltration are the most widely used industrially. Abundant examples of their application can be found in water Desalinization, the dairy industries and the separation of organic solutes from aqueous solutions. Membrane processes do not require heating, which makes the process suitable for the treatment of thermolabile products. In addition the relatively low capital and operating costs involved make membrane processes an appealing alternative to more conventional separation processes, particularly when dealing with dilute solutions.

Table: Classification of membrane separation processes.

Name of process	Driving force	Separation size range
Microfiltration	Pressure gradient	10 – 0.1μm
Ultrafiltration	Pressure gradient	<0.1 μm – 5nm
Reverse osmosis (Hyper filtration)	Pressure gradient	< 5 nm
Electro dialysis	Electric field gradient	< 5 nm
Dailysis	Concentration gradient	< 5 nm

Osmosis

Osmosis is the process of transfer of the solvent of a solution across a membrane separating two liquid solutions of different concentrations. The transfer takes place from the phase in which the chemical potential solvent is higher. A classic example is the transfer of water to a sugar solution. Figure shows two compartments separated by a water permeable membrane. One contains the solution, while the other holds pure water. Initially the level of the solution is placed at the same level as that of the water. Since the chemical potential of pure water is higher than that of water in the solution, water transfers across the membrane to the solution, increasing the level on the solution side and thus creating a hydrostatic head. When the process is at equilibrium the difference in the hydrostatic pressure between the two phases is that required to make the chemical potential of water in both phases the same. In the special case of this example, when the pure solvent is on one side of the membrane this pressure difference is the osmotic pressure, P. If solutions are at both sides of the membrane, the pressure difference gives the osmotic pressure difference ΔΠ. For very dilute solutions the osmotic pressure is given by,

$$\prod = \beta c_i RT$$

where β is a constant that takes into account the dissociation of the solute, c_i is the molar concentration of solute i, R the gas constant and T the absolute temperature. The constant (β is in general equal to the number of ions produced by the dissolved solute; thus for electrolytes that dissociate fully such as Nalco and CAC_2 it is equal to 2 and 3, respectively, while β = 1 for molecules that do not dissociate. When the solute has a large molecular mass, the osmotic pressure of the solvent ceases to depend linearly on the concentration of the solute and a polynomial approximation must be used. For concentrated solutions as well as non-ideal solutions the osmotic pressure must be determined experimentally or estimated using activities instead of concentrations.

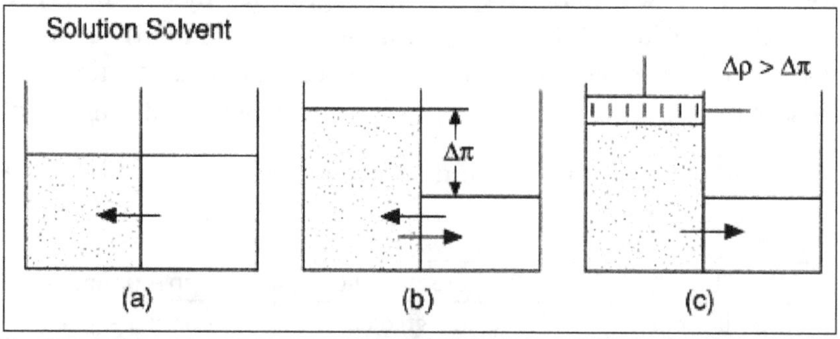

Osmosis and reverse osmosis, (a) osmotic flow, (b) osmotic equilibrium and (c) reverse osmosis.

Reverse Osmosis

The term reverse osmosis is applied to the separation of water from solutions by transfer though a semipermeable membrane that ideally should only be permeable to water. If as shown in Figure a mechanical pressure is applied to the compartment where the level is higher, the chemical potential of the solvent on the solution side will increase, forcing the transfer of the solvent through the membrane into the pure solvent phase.

This pressure difference between the phases, ΔP, creates the driving force for the transfer and is the principle on which the reverse osmosis and ultrafiltration processes are based. In order to achieve the transfer, ΔP must be greater than the difference between the osmotic pressures ΔΠ. The rate of solvent permeation across the membrane is measured as its linear velocity normal to the membrane, J_1. Ideally the membrane should be only permeable to the solvent. However this is very difficult to achieve and membranes have some permeability to the solute. The rate of permeation of the solvent is given by:

$$J_1 = K_1\left(|\Delta P| - |\Delta \Pi|\right)$$

where K_1 is the permeability coefficient of the membrane and ΔP is the transmembrane pressure. The permeability coefficient depends on the membrane resistance, the resistance to transfer caused by deposits on the membrane and the diffusion resistance in the boundary layer at the retentate side of the membrane. Depending on the thickness of the solid deposits and the hydrodynamic conditions of the retentate the two latter resistances may be negligible. In fact membrane equipment are operated under conditions that tend to minimize these two resistances in order to increase J.

The difference in solute concentration between the retentate and the permeate (in practice membranes have some permeability to the solute) creates a driving force for solute transfer across the membrane. The flux of solute, J_2, is given by,

$$J_2 = K_2\left(c_R - c_p\right)$$

where K_2 is a constant that includes the characteristics of the membrane, and c_R and c_P is the concentration of the solute in the retentate and permeate respectively.

Equations $J_1 = K_1\left(|\Delta P| - |\Delta \Pi|\right)$ and $J_2 = K_2\left(c_R - c_p\right)$ indicate that the driving forces for solvent and solute transport are independent. Therefore by increasing transmembrane pressure it is possible to increase the rate of solvent transfer without increasing the concentration of solute in the retentate. This in fact leads to an increase in the solute partition coefficient between retentate and permeate. However the increase in solute concentration in the retentate, and in particular at the membrane wall, may lead to an increase in osmotic pressure important enough to reduce J_1. This effect is called membrane polarization and will be discussed in connection with ultrafiltration.

Ultrafiltration

Ultrafiltration is also a membrane separation process driven by pressure. Its main difference from reverse osmosis is the type of membrane used. While reverse osmosis membranes are nonporous and therefore, permeable only to very small molecules, ultrafiltration membranes are microporous and permeable to small solute species in addition to the solvent. Ultrafiltration is therefore used to keep large solute species in the retentate. The effect of osmotic pressure is therefore less important in this process so that, the transmembrane pressures required are not as high as for reverse osmosis. The solvent flux obtained with ultrafiltration is also given, but the membrane permeability K_1 now depends on the characteristics of the ultrafiltration membrane and the diffusional resistances specific to this process. As a result of solvent transfer the retentate becomes more concentrated and a concentration gradient of solute builds up in the direction of transfer with the higher concentration at the membrane wall. This phenomenon, called membrane polarization, creates a concentration driving force for solute diffusion in the direction opposite to solvent transfer. At equilibrium the diffusional flux of solute back to the retentate phase equals the convective flux towards the membrane:

$$J = (c - c_p) = -D(dc/dy)$$

where J is the velocity of the permeate phase through the membrane, c_p is the concentration of solute in the permeate, D is the diffusivity of the solute in the retentate and y the distance in the direction normal to the membrane. As the separation proceeds the concentration gradient becomes steeper due to solute accumulation next to the membrane. As the solutes retained in ultrafiltration are larger than in reverse osmosis, their diffusivities are lower and thus accumulation near the membrane wall is more likely in ultrafiltration than in reverse osmosis.

Equation $J_1 = K_1(|\Delta P| - |\Delta \Pi|)$ allows comparison of the response of reverse osmosis and ultrafiltration to transmembrane pressure. As ΔP is increased, solute accumulation at the membrane wall on the retentate side is higher in ultrafiltration than in reverse osmosis due to the lower diffusivity of the solute. This accumulation of solute leads to membrane polarization as well as solute precipitation on the membrane (gelification), which in turn decrease K_1 and increase $\Delta \Pi$. The process reaches a region in which the permeate flow rate becomes independent of pressure.

Microfiltration

The separation of particles of micron and submicron levels can effectively be performed using membrane filters. The suspended particles for which the process is industrially used include colloids, microorganisms and emulsion droplets. There are two different configurations for the microfiltration operation: (1) dead-end microfiltration, and (2) crossflow microfiltration. In the first, the membrane plane is normal to the feed flux, while in the second, it is tangential. The advantages and disadvantages of the type of configuration depend on the characteristics of the feed. Highly concentrated suspensions are not suitable for dead-end treatment since the separated particles rapidly accumulate on the membrane, increasing the resistance to filtration by forming a cake and/or clogging the membrane. In crossflow filtration, the flow is parallel to the membrane and the drag forces close to the membrane wall reduce the amounts of particulate material deposited on the membrane. Therefore the first is used for dilute suspensions while the latter can deal with concentrated ones, such as slurries. Microfiltration is a pressure driven process and as such the flux of filtrate is given by Equation $J_1 = K_1\left(|\Delta P| - |\Delta \Pi|\right)$. In this case $\Delta\Pi$ can be neglected since it is not relevant to this process.

Dialysis

Dialysis is a process in which the solute, usually an electrolyte, transfers across the membrane driven by the difference in concentration between the two sides of the membrane. Two conditions have to be fulfilled for the process to be effective: (1) the concentration on the permeate side has to be kept low so that the driving force remains as high as possible, and (2) the osmotic pressure must be low, and remain low during the process, so that a counterflux of solvent does not result in feed dilution. Since the combination of these two conditions is not likely to occur, the use of selective membranes has become general practise so that the partition coefficient of the solute between the two phases can be substantially increased by using the Donnen effect. This is achieved by adding to the feed a salt with one of its ions in common with the solute and the other one rejected by the membrane. If, for example, NaCl is the solute to be separated, the addition of NaX (where X is the ion rejected by the membrane) will lead to the following expression for the NaCl partition coefficient at equilibrium:

$$\left([NaCl]_I / [NaCl]_{II}\right)^2 = 1 + [NaX]_I / [NaCl]_{II}$$

where the square brackets indicate concentrations. This equation show that the larger the concentration of NaX, the greater the separation enhancement.

Electrodialysis

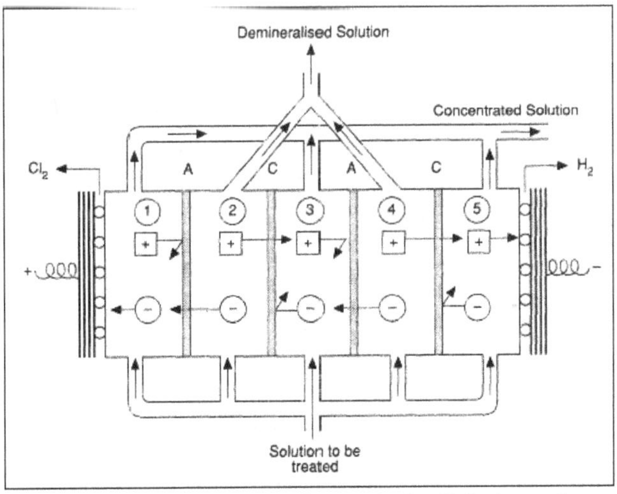

Schematic flow-diagram for an eleclrodialysis stack.

Electrodialysis is a process in which ion-selective membranes are used together with an electric field normal to the membrane phases. The basic principles of electrodialysis are better explained with reference to Figure. An electrodialysis stack consists of parallel compartments separated alternately by cation-exchange and anion-exchange membranes. The feed is introduced into each compartment, so that aided by the electric field and selected by the membranes, the cations and the anions are transferred in opposite directions to the neighboring compartments. In this way a demineralized solution leaves compartments 2 and 4 and the concentrated one compartments 1, 3 and 5.

Depending on the type of membrane used, ions can be selected according to their valence. Therefore this process can be used for fractionating ions of different valence. The main application of electrodialysis is in demineralization in general and in the desalinization of brackish water.

Pervaporation

Pervaporation is the combination of the selective separation and transfer of a component across the membrane and its evaporation on the permeate side. In order to achieve evaporation the pressure on the permeate side must be such that the partial pressure of this component is lower than its saturation vapours pressure. Therefore although the driving force for transfer is the difference in activity of the transferred species, this is the result of applying a vacuum on the permeate side.

Pervaporation is more expensive than other membrane processes due to the required supply of heat to produce the evaporation. It is, therefore, used in the separation of components from mixtures that are difficult to treat, such as azeotopic solutions and isomer mixtures, for which conventional processes would be more costly.

Liquid Membranes

A liquid membrane is a liquid phase that separates two fluid phases of different composition. In most applications the liquid membrane is an organic phase placed between two aqueous phases. The principals involved in liquid membrane separations are those of solvent extraction, rather than membrane separation. The solute is first selectively extracted by the liquid membrane, it then transfers across the liquid film driven by its concentration gradient and when it reaches the other side of the liquid membrane it is stripped by the third phase.

There are two types of mechanisms of transfer to and from the membrane: (1) Physical transfer and (2) carrier mediated transfer. The first is based on the solubility of the solute in the membrane, while the second requires the presence of a selective reactant in the organic phase called the carrier. In this case the process is based on the reversibility of the chemical reaction. The extraction of Zn^{2+} from an aqueous feed with a carrier RH dissolved in the organic membrane will be used as an example of carrier mediated extraction. The overall reaction between the cation and the carrier can be represented by,

$$Zn^{2+} + 3RH \Leftrightarrow ZnR_2HR + 2H^+$$

where bars indicate species in the organic phase. Reaction is reversible. At pH 3 the forward reaction takes place while at low pH values the reverse reaction predominates. A suitable liquid membrane system for the separation of a membrane containing RH should have a high pH value in the feed phase and a low one in the stripping phase. Figure shows a schematic drawing of the concentration gradients in the feed, the membrane and the stripping phase with the corresponding concentration profiles.

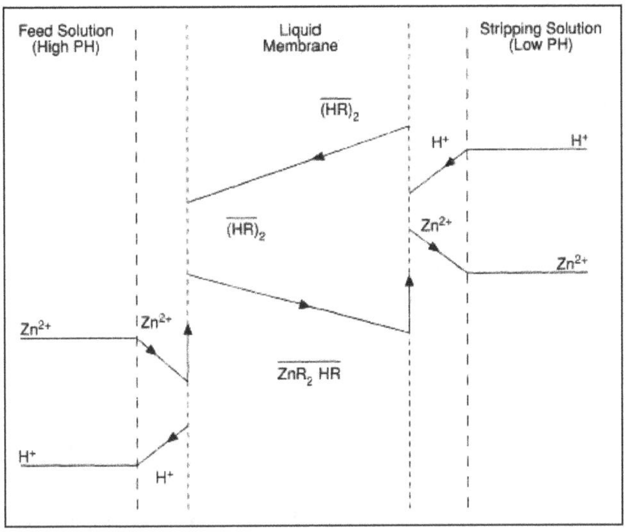

Concentration profiles in the extraction of zinc with a liquid membrane.

The main advantage of liquid membranes over conventional solvent extraction is that by contacting the three phases simultaneously, the solute partition coefficient between the two aqueous phases may be several orders of magnitude greater than the one that can be obtained with conventional solvent extraction.

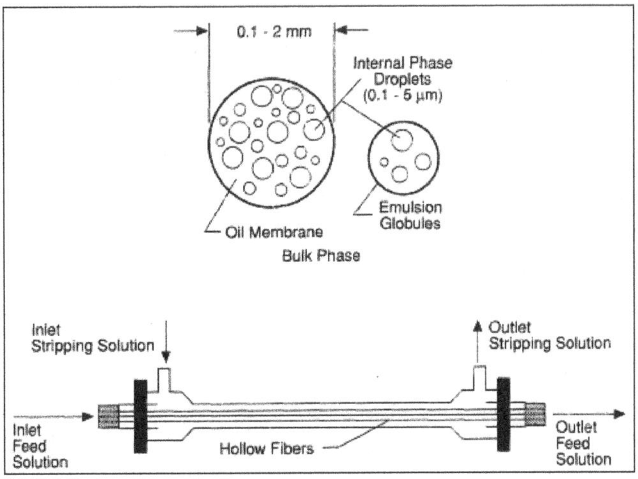

Liquid membranes: (a) emulsion liquid membrane, (b) supported liquid membrane.

There are two ways in which a liquid membrane can be formed leading to two configurations: (1) emulsion liquid membranes and (2) supported liquid membranes. These are illustrated in Figures. The emulsion liquid membrane is obtained by adding a surfactant into the organic phase in order to stabilize a water in oil emulsion. The emulsion is then dispersed in the feed so that the organic phase, which is the continuous phase of the dispersed globules, becomes the membrane. Supported liquid membranes are obtained by impregnation of porous fibers of hollow fiber modules.

Adsorption

Adsorption is the process by which molecules of a substance from a gas mixture or liquid solution became attached to a solid or liquid surface. The substance being absorbed is an adsorbate and the absorbing substance, an adsorbent. Adsorption is originated by the surface forces acting on the solid-gas, solid-liquid, gas-liquid or (in the case of immiscible liquids in contact) the liquid-liquid interface.

Adsorption of Gases by Solids

In adsorption of gas by a solid surface, one never distinguishes between physical adsorption and chemisorption. In physical adsorption, the main factors are the nature of interacting substances, the forces of interaction on the interface, and the interaction between adsorbate molecules. Of paramount importance is the degree of surface inhomogeneity and the capability of adsorbed molecules to be in translational, rotary, and oscillatory motion. The surface structure and, hence, adsorption properties may change during adsorption. This is most clearly pronounced when the so-called adsorption energy is of the same order as the surface energy of the adsorbent, e.g., ice, paraffin, and polymers.

Equilibrium in physical adsorption is rapidly established and is reversible. It should be born in mind, however, that the actual adsorption rate can be restricted by the mass transfer rate in the gaseous phase or inside the pores of a porous adsorbent, i.e., by the adsorbate supply to the surface. It is believed that physical adsorption is brought about by the same intermolecular forces as vapor condensation. Hence, a conclusion is drawn that the heat of physical adsorption is close to that of condensation and the amount of physical adsorption is particularly high when the gas has a temperature below critical, i.e., in a vapor.

Gas adsorption by a solid surface is expressed quantitatively as a gas volume adsorbed by the adsorbent per unit mass or unit surface of the adsorbent. Chemisorption, in contrast to physical adsorption, involves chemical reactions on the surface, which may result in its rearrangement. Cases are also known where physical adsorption proceeds over a chemisorbed layer.

The parameter describing the physical adsorption dynamics is adsorption time, which is the mean lifetime of the molecule on the adsorbent surface. It is $t = t_0 \exp Q/RT$, where t_0 is the time of molecular vibration ($t_0 = 10^{-13} - 10^{-12}$ s) and Q is the energy of the molecule-surface interaction, i.e., the heat of adsorption. If thermal equilibrium is established between the molecules and the surface, then the molecules being desorbed leave the surface in the direction independent of the incidence direction, i.e., the accommodation coefficient is assumed to be unity).

An important characteristic of adsorption is also a molecule concentration on an adsorbent surface. It is measured in mol/cm² and calculated as $\Gamma = zt$, where z is the number moles of gas colliding with unit surface in a second. At a given temperature T, adsorption can be classified according to the parameters shown in Table.

Table: Regions of adsorption.

Q, kcal/mol	t, s	Γ, absorbate concentration, mol/cm_2	
0.1	10^{-13}	0	No adsorption, specular reflection of molecules, the accommodation coefficient is zero.
1.5	10^{-12}	0	The region of physical adsorption, the accommodation coefficient is one.
3.5	10^{-11}	10^{-12}	
9.0	10^{-7}	10^{-8}	
20	100		The region of chemisorption
40.0	10^{17}		

An isotherm, i.e., the dependence graphed as a function $V = f(p)$, where V is the volume adsorbed and p the pressure, is the most common representation of adsorption. In 1918 Langmuir derived theoretically an adsorption isotherm equation that became known as the Langmuir equation. He assumed that the adsorbent surface possesses a certain number of adsorption sites s, part of which are occupied by adsorbed molecules s_1 and the other part are vacant s_2. He considered adsorption-desorption as a process of condensation and evaporation. The rate of condensation was taken to be proportional to the number of free surface sites s_2, i.e., $k_2 s_2$, while the rate of evaporation proportional to the number of occupied surface sites, i.e., $k_1 s_1$. When the equilibrium was attained $k_1 s_1 = k_2 p s_2 = k_2 p(s-s_1)$ or $\theta = bp/(1+b)$, where p is the partial pressure of an adsorbate, $\theta = s_1/s$ is the fraction of the surface occupied by adsorbed molecules and b is the Langmuir constant. Taking into account the fact that the evaporation rate is proportional to the saturation vapor pressure p_G, the Langmuir adsorption isotherm equation can be reduced to $\theta = (cx)/(1 + cx)$, where $c = \exp Q/RT$ and $x = (p/p_G)$ is the ratio of the vapor pressures of a given substance in the mixture to its saturation vapor pressure at the same temperature T. The heat of adsorption Q can be assumed to be equal to the extent of heat of vaporization per mole. R is the universal gas constant related to a mole of vapor. The next step in the description of adsorption isotherm is the allowance for formation of multi-molecular layer, which was realized in the BET (Brunauer-Emmet-Teller) equation:

$$\frac{V}{V_m} = \frac{cx}{(1-x)\left[1+(c-1)x\right]}$$

where V and V_m are respectively the actually adsorbed gas volume and its limit in a monolayer. This equation has gained recognition as a most reliable adsorption isotherm equation and has been widely employed for determining the surface per unit volume from the experimental data. It can be transformed to the form,

$$\frac{x}{V(1-x)} = \frac{1}{cV_m} - \frac{(c-1)x}{cV_m}$$

Plotting x/V(x – 1) against x gives a straight line the intercept and slope of which is used to find the constants V_m and c. The diversity of experimental adsorption isotherms is depicted in Figure. Type 1 isotherms correspond to the Langmuir equation and are characterized by a monotonic approach to a limit value conforming to a complete monolayer surface coverage by a monolayer of adsorbed molecules. Type 2 corresponds to the case when first, a monolayer is formed, and then (from the point B) a multi-molecular layer predominantly develops. Isotherms of Type 3 are relatively uncommon and correspond to the formation of multi-molecular layer when heat of adsorption is less than or equal to adsorbate heat of condensation. Types 4 and 5, as a rule, are characteristic of capillary adsorbate condensation in porous solids.

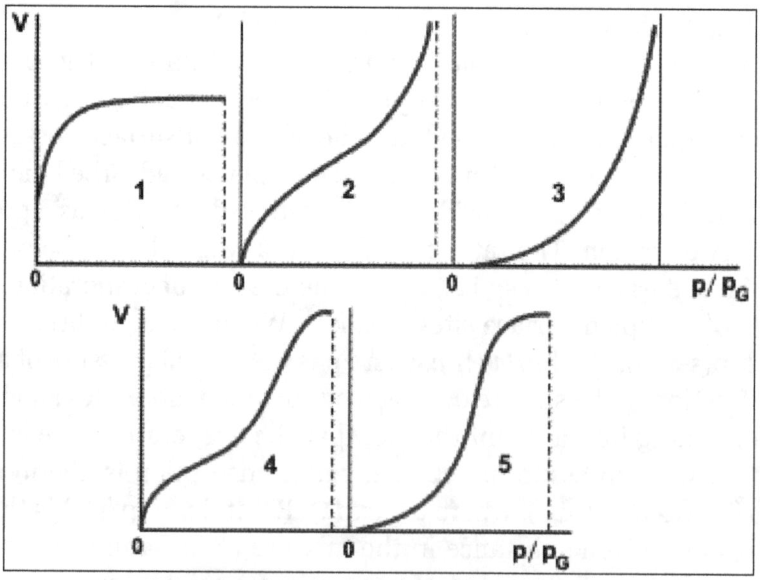

Forms of experimental adsorption isotherms.

Adsorption in porous solids with pores substantially exceeding the free path length of molecules is of the same nature as it is on an open surface of a nonporous solid. In particular, at temperatures higher than the critical temperature, no multi-molecular layers are formed. An adsorbent with molecular size pores behaves like a Molecular Sieve, and its apparent surface per unit volume depends on the size of adsorbed molecules: small molecules pass into pores and adsorb on their surfaces, while large ones are trapped and no adsorption occurs in pores.

Of great interest are zeolites composed of tetrahedral groups (Al, Si) O_4. They have large voids into which the molecules of adsorbed gas have to pass through much smaller holes. Thus, in chabazite Ca $Al_2Si_4O_{12}$, there are six inlet holes about 4 Å in diameter in the void 10 Å in diameter. These holes accommodate only monatomic or diatomic gases, H_2O, and H-alkanes, but trap larger molecules. Changing the hole size makes it possible to separate gases.

At temperatures below critical, for multi-molecular adsorption in small size pores not only is the number of layers restricted but also so-called capillary condensation may be observed. For capillary condensation to occur, the effective curvature of the meniscus in the micropores must be sufficient to entirely fill the pores with condensing liquid; however, since the curvature is variable even with a narrow range of pore size, the desorption and adsorption curves do not coincide leading to a hysteresis in adsorption.

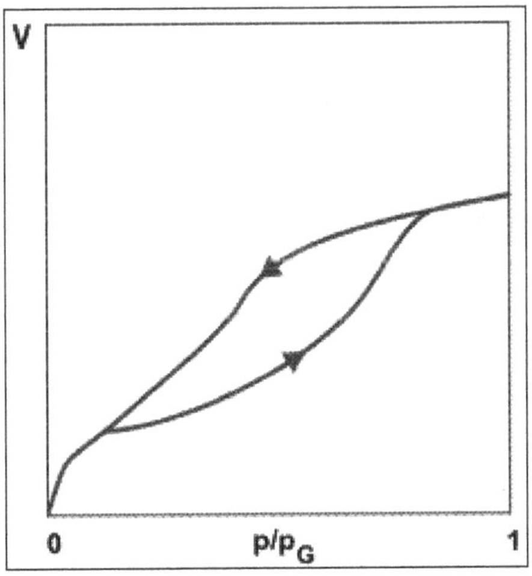

Adsorption isotherm hysteresis due to capillary condensation.

Chemisorption gives rise to bonds between adsorbate and adsorbent approaching chemical bonds in strength. In this case, the chemical properties of adsorbate and adsorbent may differ from their initial ones. Chemisorption conspicuously shows up in formation of the first molecular layer because in this case, an enhanced heat of adsorption characteristic of chemisorption comes into play. Considerable variation may occur because of inhomogeneities in the surface or in the energy distribution of adsorption centers. The inhomogeneity may be attributed to surface defects and impurity inclusions, i.e., it depends on the material treatment and quality.

As the second and subsequent molecular layers are formed, chemisorption is reduced to physical adsorption. The distinguishing feature of chemisorption, in contrast to physical adsorption, is the fact that it can proceed with gases, whose temperature is above critical. The rate of chemisorption is described by the expression:

$$R \propto = d\theta / d\tau = k_2 f(\theta) p$$

Where $f(\theta)$ is a function depending on the nature of adsorption; k_2, the constant of adsorption rate; $k_2 = k \exp\left(-E_{\partial}^{\bullet} / RT\right)$; and $E_{\partial}^{\bullet}$ is the activation energy which, taking into account its dependence on the surface coverage θ, is described by the binomial $E_{\partial}^{\bullet} = E_{\partial}^{\bullet} + \alpha\theta$. Hence, at a given temperature, the rate of adsorption varies with surface coverage by molecules in proportion to $f(\theta) \exp(-\alpha\theta)$.

The activation energies of desorption and adsorption differ by the heat of adsorption $Q: E_d^{\bullet} = E_{\partial}^{\bullet} + Q$. The rate of desorption is described as $R_d = d\theta/d\tau = kf'(\theta) \exp (E_{\partial}^{\bullet} + Q/RT)$, where $f'(\theta)$ is analogous to $f(\theta)$. Chemisorption is an important underlying phenomenon in heterogeneous catalysis.

Adsorption from Solutions

In adsorption from solutions, the behavior depends greatly on whether the solutions are of nonelectrolytes or electrolytes. Adsorption from nonelectrolyte solutions depends on adsorbate concentration and, in the case of dilute solutions, is similar to gas adsorption. The solvent properties manifest themselves at high concentrations.

Two basic models of nonelectrolyte adsorption exist. Adsorption in terms of the first model is restricted to a monolayer directly on the surface, whereas the consecutive layers are merely an ordinary solution. This resembles gas chemisorption, but the heat of adsorption from a solution is low in relation to the energy of chemical reaction. In the second model, adsorption occurs in a fairly thick (up to 100 Å) multi-molecular interphase layer in a decreasing potential field of the solid surface.

Both models are consistent with experiment and yield the same mathematical expressions, e.g., an adsorption isotherm: $\theta/ (1 - \theta) = ka_2/a_1$, where a_1 and a_2 are activities of the solvent and the solute in the solution. This is obviously an expression similar to the Langmuir isotherm equation.

Adsorption in electrolyte solutions is considered using several approaches. If an electrolyte is assumed to be adsorbed as an entity bound by opposite charged ions, then this resembles adsorption in nonelectrolyte solutions. However, more frequently, the ions of one sign are more strongly attracted by the surface than the ions of the other sign. The latter produce near the surface a diffuse layer through which the adsorbed ions are forced to diffuse.

Adsorbers

A wide range of adsorbents are used for adsorption from gases and liquids. Special adsorbents are often developed for specific purposes. Among the most well-known adsorbents are activated carbon used for adsorbing gases, including hydrocarbons, and silica gel for adsorbing water vapor.

A technological apparatus in which adsorption is implemented is called an adsorber. Adsorbers can have an immovable adsorbent bed and operate either with periodic regeneration under an adsorption-desorption regime or without regeneration but with periodic change of the adsorbent.

In fixed-bed adsorbers, the adsorption process is represented by the breakthrough curve. The concentration of pollutant C in the exit gas stream is plotted versus time. Initially, the adsorbent adsorbs the pollutant gas from the stream readily and efficiently, so that the pollutant concentration of the outlet stream is close to zero (period 0-a on the abscissa in. In this period, the bulk of the adsorption is taking place near the inlet with the rest of the bed removing the traces. Eventually, the inlet region becomes saturated and the main region of adsorption moves towards the outlet. Traces of the adsorbate begin to appear at the outlet. When the outlet concentration begins to rise rapidly (B), the so-called "break through point" has been reached. Though, at this time, the exit pollutant concentration may be below the required emission standard C_{std} after the point of breakthrough, the exit concentration rises rapidly towards the inlet concentration C_0. Thus, at time B, regeneration is required. A steep breakthrough curve is more desirable than a flat one. For a steep curve, the bed saturation may reach 80 per cent, but with a flat curve only 15-20 per cent may be typical before breakthrough. The calculation methods for adsorption, particularly in porous adsorbents, mist allow for heat and mass transfer to the adsorbed substance in pores.

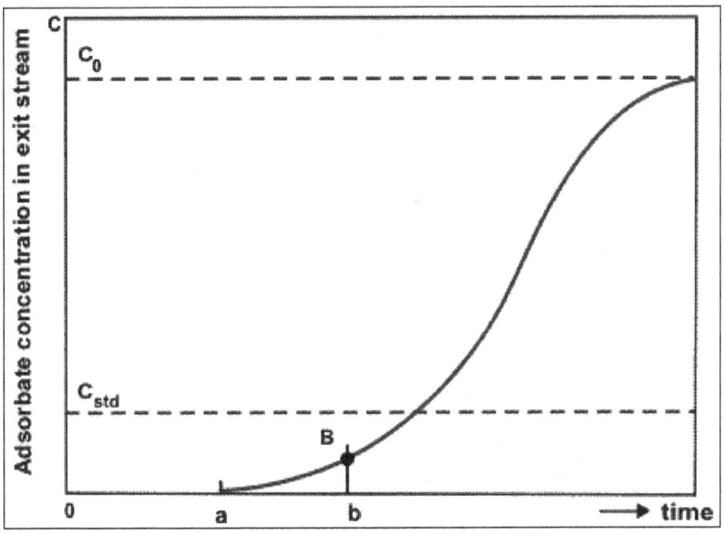

The breakthrough curve.

Crystallization Process

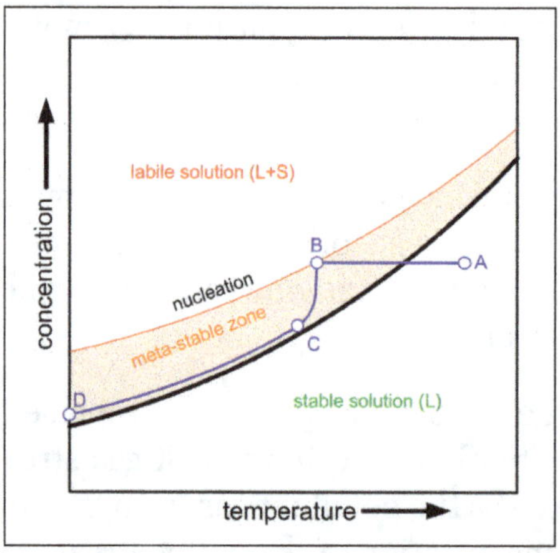

Solubility diagram and typical de-supersaturation profile of an unseeded crystallization.

Theoretical Background

Crystallization is the process of forming a crystalline material from a liquid, gas or amorphous solid. The crystals thus formed have highly regular internal structure, the basis of which is called the crystal lattice. Since the formation of such a highly ordered structure prohibits foreign molecules from being incorporated into the lattice, a solid product of high purity is obtained. The simultaneous formation and purification of a solid product makes crystallization an important operation in the process industry.

All crystallization processes are aimed at creating a supersaturated solution or melt. The supersaturation is the driving force under whose influence new crystals are formed and present crystals grow. This is illustrated in figure. Where a typical cooling crystallization experiment in shown together with the solubility line.

For most substances, the solubility increases with increasing temperature, with sodium chloride being a notable exception. Suppose we start at point A in the diagram, which is undersaturated. Any crystals added to a solution in this region would dissolve. If we now cool to a point between A and B, we enter the meta stable region where existing crystals will grow, but no new crystals are formed. Cooling further we obtain a labile solution at point B where spontaneous formation of new crystals, i.e. nucleation, takes place. This dramatically decreases the concentration and point C will be reached. Cooling further, the crystals formed between B and C grow and consume whatever supersaturation we create by cooling, so we stay in the meta stable region until we reach the end of the crystallization at point D. The important phenomena that occur during crystallization can be described by the following quantities:

- The nucleation rate $B\left[\#/\left(m^3 s\right)\right]$; the number of new crystals formed per unit time and volume of suspension.

- The growth rate G [m/s]; the rate at which the size of the crystals increases.
- Both B and G are influenced by the supersaturation S [-], where S = c/cs, with c [kmol/m³] the current solute concentration and cs the equilibrium concentration at the same temperature.

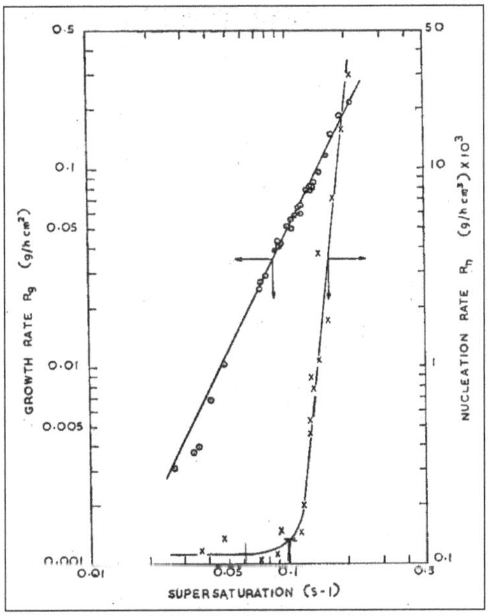

Growth and nucleation rates of potassium sulphate in water as a function of supersaturation.

For potassium sulphate, the dependence of B and G on S is shown in figure. A further differentiation of nucleation can be made by the way new crystals are formed. Secondary nucleation can take place at low supersaturations when crystals are already present. Small parts of the existing crystals are washed off and form new nuclei that can grow. As such, secondary nucleation is severely affected by the stirring rate. Primary nucleation usually takes place in the absence of previously formed crystals (though this is not a necessity) when solute molecules meet and form clusters that can grow further to form crystals. This is a stochastic process, and as such, nucleation can be somewhat unpredictable.

The formation of stable nuclei only happens if the increase in interfacial energy resulting from the formation of a solid–liquid interface is offset by a decrease in energy released by the formation of a (thermodynamically favored) solid. At low supersaturations, the energy required to form a solid–liquid interface is bigger than the energy released upon formation of a cluster which are consequently unstable resulting in a very low nucleation rate. Only when the supersaturation is high enough, i.e. when we enter the labile region, stable solid particles will form. This is generally a very sharp transition, which can also be seen from the steep increase in nucleation rate for a certain supersaturation in figure.

The unpredictable nature of nucleation, coupled with its dramatic effect on the crystal size distribution, indicate that this a process we can poorly control. As such, in industrial crystallization, we mostly add small, previously produced crystals. These crystals will consume whatever supersaturation we create by growing, thus ensuring we stay within the meta stable region. This is called seeded crystallization, a technique often applied if a consistent product quality is important.

In cooling crystallization, the supersaturation is created by a decrease in temperature. Depending on how the solution is cooled, a different end product is obtained. The simplest way of performing a cooling crystallization is to let a warm solution cool to the surrounding room temperature. Obviously, the cooling rate will be highest in the beginning, when there is a large temperature difference between the solution and its surroundings, and it will be lowest at the end of the crystallization. The seed crystals will not be able to grow fast enough to consume the supersaturation created in the beginning. As a consequence, we are likely to shoot into the labile region and have nucleation. Therefore, we can expect our end product have a rather wide distribution of crystal sizes.

Alternatively, we can cool linearly with the help of a thermostat (curve ii in figure). This will reduce the initial supersaturation, but since the solubility curve is steepest at high temperature, we will still create the highest supersaturation in the beginning we therefore still risk having nucleation.

A more sophisticated cooling would ensure that the supersaturation remains constant throughout the whole crystallization. This requires knowledge on the growth rate, the nucleation rate, the solubility and the seed crystals, but with some mathematical effort, it is possible to compute a cooling curve that results in a constant supersaturation. In figure, the evolution of the supersaturation for each of these cooling strategies is shown.

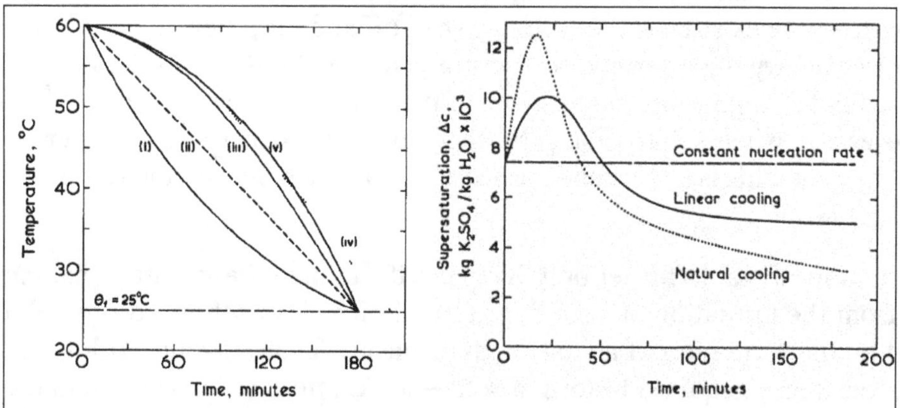

(a) (b)

Different cooling curves (a) and the resulting evolution of the supersaturation as a function of process time (b).

To calculate the cooling curve for constant supersaturation, the following assumptions are made:

- We perform a batch crystallization.

- The supersaturation S is constant throughout the whole process.
- The supersaturation S is small enough to prevent nucleation from happening.
- The seed crystals are mono disperse, i.e. all seed crystals have the same size.

The balance equation for such a process is as follows:

$$\frac{d\Delta c}{dt} = r - k_g A \Delta c^n - k_n \Delta c^m$$

where $\Delta c = c - c_s$ is the difference between actual and saturation concentration (sometimes called supersaturation as well) and r is the increase in the supersaturation due to the decreasing temperature. The second term of the right hand side is the decrease of the supersaturation due to crystal growth and the third term is the decrease due to nucleation. The crystal area per unit mass of solvent can, if nucleation is neglected (k_n=0), be calculated as follows:

$$A(t) = \beta N L(t)^2$$

where β is an area shape factor and L(t) is the crystal size:

$$L(t) = L_0 + \int_0^t G(\theta(t), L(t))dt$$

N is the number of seed crystals:

$$N = \frac{W_0}{\alpha \rho L_0^3}$$

W_0 is the mass of seed crystals per unit mass solvent and α is a volume shape factor. The crystal growth rate can be described as follows:

$$G(\theta(t), L(t)) = \frac{kg\beta}{3\alpha\rho}\Delta c^n$$

Since for r, we can write:

$$r = \frac{d\Delta c_s}{dt} = \frac{d\Delta c_s}{d\theta}\frac{d\theta}{dt}$$

The balance equation can be simplified by assuming $\frac{d\Delta c}{dt} = 0$.

$$\frac{d\theta(t)}{dt} = \frac{3w_0 G(\theta(t), L(t))}{\frac{dc_s}{d\theta}L_0}\left(1 + \frac{1}{L_0}\int_0^t G(\theta(t), L(t))dt\right)^2$$

This general equation can now be applied to the potassium sulphate system. This requires certain substance specific data to be known. The measurement of these is often very difficult and extremely time consuming.

Nitration

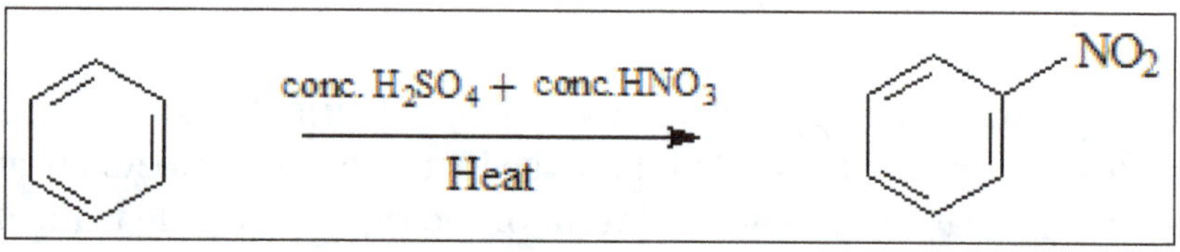

Substitution of a nitro group into the aromatic ring is nitration. It is an aromatic substitution reaction. It is an electrophilic substitution reaction. Substitution takes place involving an electrophile. Similar reactions are: Sulphonation, Halogenation, Alkylation and Acylation.

The Mechanism

Generation of Electrophile

$$2H_2SO_4 + HNO_3 \rightleftharpoons NO_2^{\oplus} + H_3^{\oplus} + 2HSO_4^{\ominus}$$

- Step 1: Formation of σ complex.

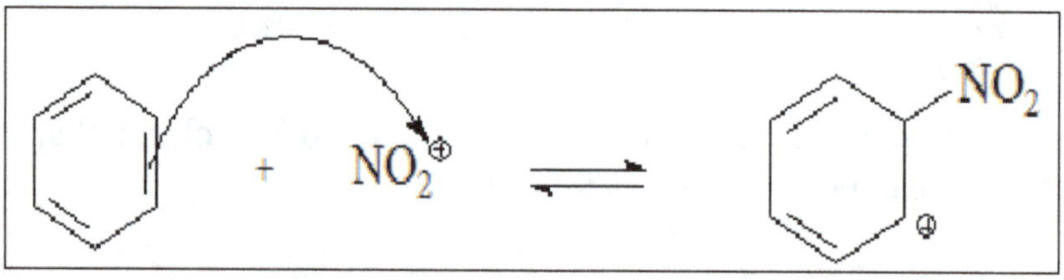

- Step 2: Re-aromatisation.

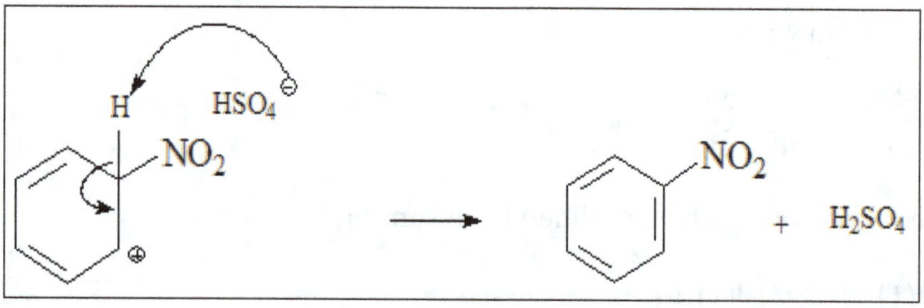

Interaction of sulphuric and nitric acids results in the electrophile nitronium ion NO_2^+. The Pi-electrons of benzene ring attack the electrophile, forming a σ complex.

Deprotonation of the σ complex, by the base yields the product. The σ complex is also known as wheland intermediate or arenium ion. It is a two-step process.

The first step is slow and reversible. It is slow because bond breaking is involved; hence there is an activation energy barrier. A stable aromatic compound is becoming a fairly unstable product. It is reversible because the products in the step can easily recombine to go back to the reactants (very little activation energy barrier).

The second is the fast step (formation of a highly stable product from an unstable reactant, rearomatisation). The base HSO_4^- deprotonates the complex in the second step. H_2SO_4 donates a proton to HNO_3 hence it is an acid and HNO_3 is a base in comparison in this specific reaction.

Generation of the electrophile some more details:

$$\begin{array}{llllllll} H_2SO_4 & + & HNO_3 & \rightarrow & HSO_4^{\ominus} & + & H_2NO_3^{\oplus} & \\ H_2NO_3^{\oplus} & & & \rightarrow & H_2O & + & NO_2^{\oplus} & \\ H_2SO_4 & + & H_2O & \rightarrow & HSO_4^{\ominus} & + & H_3O^{\oplus} & \\ 2H_2SO_4 & & +HNO_3 & \rightleftharpoons & NO_2^{\oplus} & + & H_3O^{\oplus} & 2HSO_4^{\ominus} \end{array}$$

$$H\ddot{O}\text{-}NO_2 \xrightleftharpoons{H_2SO_4} H_2\overset{\oplus}{O}\text{-}NO_2 + HSO_4^{\ominus} \xrightleftharpoons{H_2SO_4} H_3O^{\oplus} + HSO_4^{\ominus} + NO_2^{\oplus}$$

The electrophile Nironium ion is generated as shown.

- Nitric acid gets protonated, in a medium such as sulphuric acid to form $H_2O+.NO_2$ which looses water to form nitronium ion. The mixture thus has four ions.
- The function of sulphuric acid is not as a reagent for absorbing the water that is formed.
- Solution of nitric acid in pure sulphuric acid shows a fourfold depression in freezing point which can be interpreted as due to the presence of four ions.
- Nitration is very slow in the absence of sulphuric acid. Sulphuric acid itself has no influence on benzene under the conditions of the reaction. So sulphuric acid is acting on nitric acid rather than on benzene.
- The electrophile in the reaction is NO_2^+, why not H_3O^+.
- Though Oxygen atom in H_3O^+ has a positive charge it is still not an electrophilic centre, because the atom has a complete octet in its outermost shell hence cannot accommodate any more electrons.

Nitronium ion as an Electrophile

The presence of NO_2^+ in solution and as part of a salt has been confirmed. Nitronium perchlorate ($NO_2^+ClO_4^-$), Nitronium fluoroborate ($NO_2^+BF_4^-$), have been prepared and are found to be effective nitrating agents at even sub-zero temperatures.

A line in the Raman spectrum at 1400 cm^{-1}, corresponds to a linear triatomic species. The function of sulphuric acid is to just generate the nitronium ion by interacting with nitric acid, if that is so any other strong acid should promote nitration. It has been found to be so, $HClO_4^-$ and HF + BF_3 are also effective. The actual isolation of σ complex in some processes lends credibility to the mechanism as in the following case:

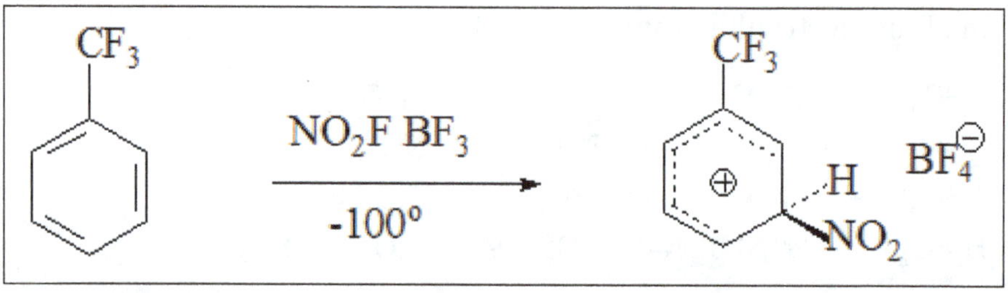

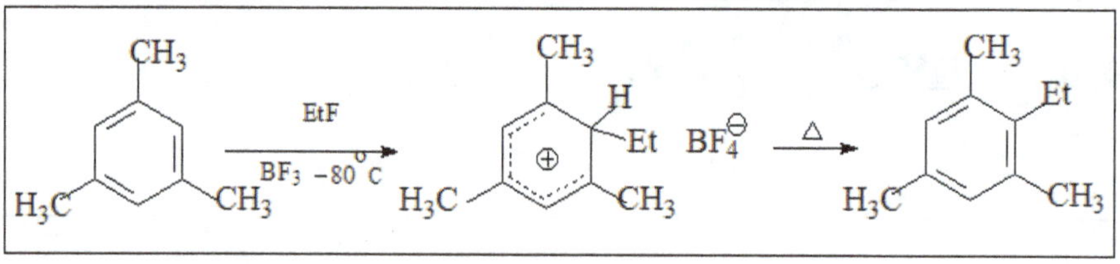

In the second step HSO_4^- depotonates the σ complex leading to nitrobenzene. The σ complex is unstable by loss of a proton it can form a very stable aromatic species. This happens when a base from the medium is available.

Nitration is not a Single Step Process

Nitration is carried out using C_6H_6 and C_6D_6 under identical conditions and the rate is measured in each case (kH , kD). If the C-H or C-D bond is breaking in the rate determining step the two rate constants should be different k_D should be much higher than k_H, (about 7). In actual fact they have been found to be almost similar in value. Therefore the C-H or C-D bond is not breaking in the rate determining step that is it is breaking in another step. Conclusion is that the mechanism is not a single step process.

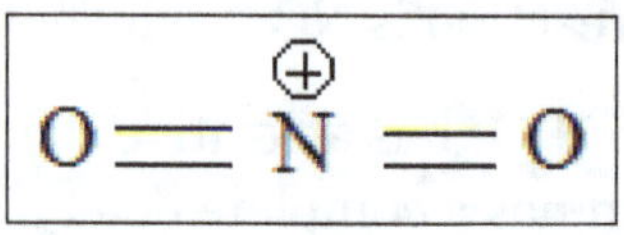

Structure of Nitronium ion.

The species is linear-bond angle: 180°. Distance between Nitrogen and oxygen: 115pm.

The sigma complex is resonance stabilised.

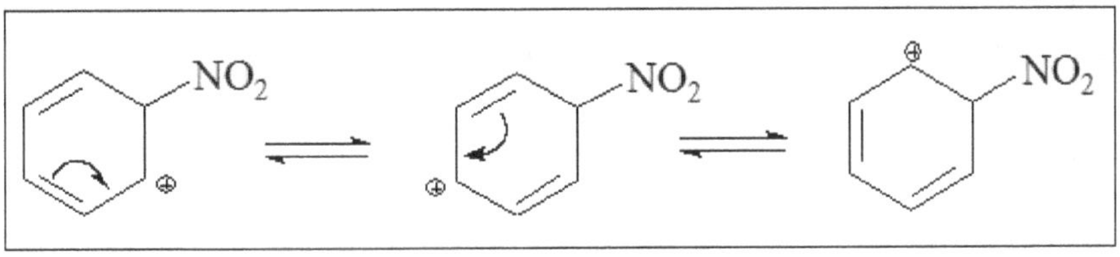

In the generation of electrophile four ions are involved, how do you know? The combination of sulphuric and nitric acids is called the nitration mixture.

Halogenation

Halogenation is a chemical reaction that involves the reaction of a compound, usually an organic compound, with a halogen. The pathway and stoichiometry of halogenation depends on the structural features and functional groups of the organic substrate as well as the halogen. Example: Alkenes readily react with halogens under standard laboratory conditions.

Halogens Facts:

- Chlorine, fluorine, bromine, iodine are all halogens.
- All these halogens have seven electrons in their outer orbit, i.e. they are monovalent.
- Halogens have lower melting and boiling points compared to other non-metals.
- All halogens are poor thermal conductors and also poor conductors of electricity.
- Except fluorine, all other halogen elements can expand their shell to include valence electrons. They can hold around fourteen valence electrons in their outer shell.

In this addition reaction, halogens atoms are added across the double bond of the alkene. In an alkene containing only one double bond, the double bond is broken, the halogen atoms are added, and, the only product of the reaction will be a dihaloalkane. In a bromination reaction, if excess alkene is present, then the reaction mixture will change from a red-brown colour to colourless.

Ethene + Bromine Water → (UV) Bromoethane + Hydrogen bromide

There are several processes for the halogenation of organic compounds, including free radical halogenation, ketone halogenation, electrophilic halogenation, and halogen addition reaction. The determining factors are the functional groups. Saturated hydrocarbons typically do not add halogens but undergo free radical halogenation, involving substitution of hydrogen atoms by halogen. Unsaturated compounds, especially alkenes and alkynes, add halogens. Aromatic compounds are subject to electrophilic halogenation.

The facility of halogenation is influenced by the halogen. Fluorine and chlorine are more electronegative and are more aggressive halogenating agents. Bromine is a weaker halogenating agent than both fluorine and chlorine, while iodine is least reactive of them all. The facility of dehydrohalogenation follows the reverse trend: iodine is most easily removed from organic compounds and organofluorine compounds are highly stable. Bromination and iodination are more likely to substitute at the beta carbon.

Halogenation reactions with elemental chlorine, bromine, and iodine are of considerable importance. Because of high exothermocities, fluorinations with elemental fluorine tend to have high levels of side reactions. Consequently, elemental fluorine is generally not suitable for direct fluorination. Two types of reactions are possible with thesehalogen elements, substitution and addition.

Substitution halogenation is characterized by the substitution of a halogen atom for another atom (often a hydrogen atom) or group of atoms (or functional group) on paraffinic, olefinic, aromatic, and other hydrocarbons. A chlorination reaction of importance that involves substitution is that between methane and chlorine.

Addition halogenation involves a halogen reacting with an unsaturated hydrocarbon. Chlorine, bromine, and iodine react readily with most olefins; the reaction between ethylene and chlorine to form 1, 2-dichloroethane is of considerable commercial importance, since it is used in the manufacture of vinyl chloride.

Addition reactions with bromine or iodine are frequently used to measure quantitatively the number of CH&dbnd;CH(or ethylenic-type) bonds in organic compounds. Bromine numbers or iodine values are measures of the degree of unsaturation of the hydrocarbons.

Substitution halogenation on the aromatic ring can be made to occur via ionic reactions. The chlorination reactions with elemental chlorine are similar to those used for addition chlorination of olefins.

Chlorinated paraffins have been widely used in industry throughout the world since their introduction nearly 40 years ago. Chlorinated paraffins are needed where chemical stability is highly desired. The primary application of Chlorinated paraffins is in industrial cutting fluids, particularly in the manufacture of automobiles and automobile parts. In addition to their use in cutting oils, Chlorinated paraffins are also used in a lot of commercial paints, adhesives, sealant and caulks.

Chlorinated paraffins are a family of complex substances representing more than 200 commercial products. Use applications for chlorinated paraffins range from extreme pressure additives in lubricants, to secondary plasticizers in paints and plastics, to flame retardants in various plastics and textiles. Chlorinated paraffins are produced through chlorination of straight chain paraffin fractions which are typically subdivided into three categories based on their carbon chain lengths. The properties of the different chlorinated paraffins can vary significantly depending on their carbon chain length and degree of chlorination.

The conventional methods of halogenation of organic compounds caused serious Environmental problems. Further these methods are expensive too. New reagent system may have commercial importance by saving resources including time, health, and environment and may be cost effective.

Generation of Free Radicals

Three general methods are used for the generation of free radicals.

Thermal Generation

Two types of compounds dissociate to give free-radicals at moderate temperature: (i) compounds that have an intrinsically weak bond such as dialkyl peroixides ($D_{O\text{-}O}$ = 155 KJ mol^{-1}), and (ii) compounds that, on fragmentation, form strongly bonded products, such as AIBN which releases N_2.

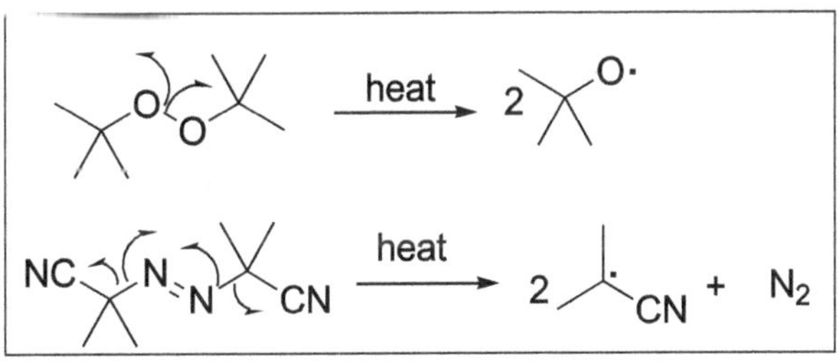

Photochemical Generation

Light can bring the fragmentation of a compound if the wavelength of the light is correspond both to an energy greater than that of the bond to be cleaved, and to an electronic excitation of the molecule concerned. This procedure is suitable for the formation of alkoxy radicals from alkyl nitrite or hypochlorites.

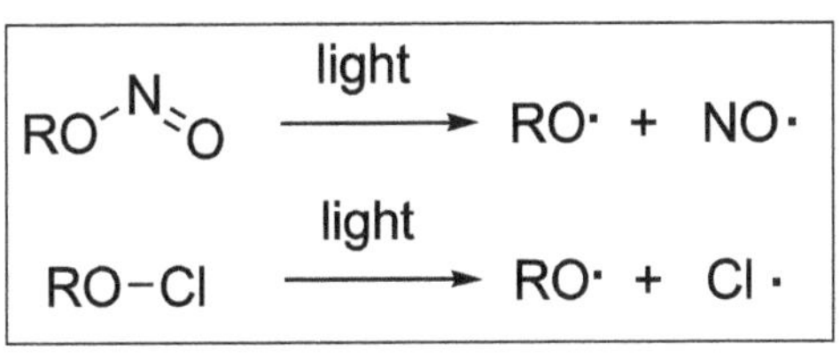

Redox Generation

Covalent bonds may be broken by electron transfer process either by accepting an electron from a donor or donating an electron to an acceptor.

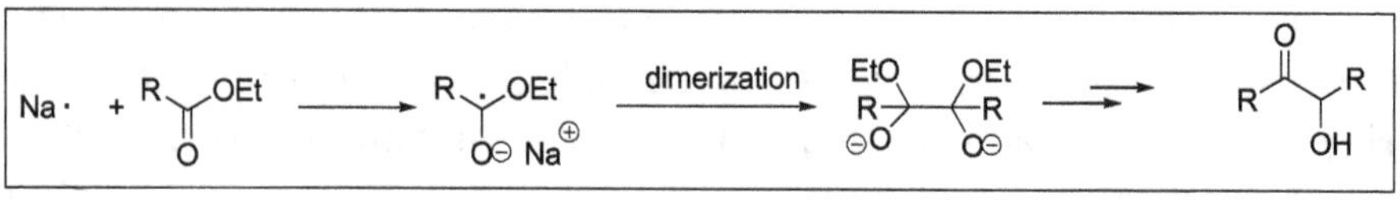

In the second group, the acceptor is generally a transition-metal-ion in a high-valence state.

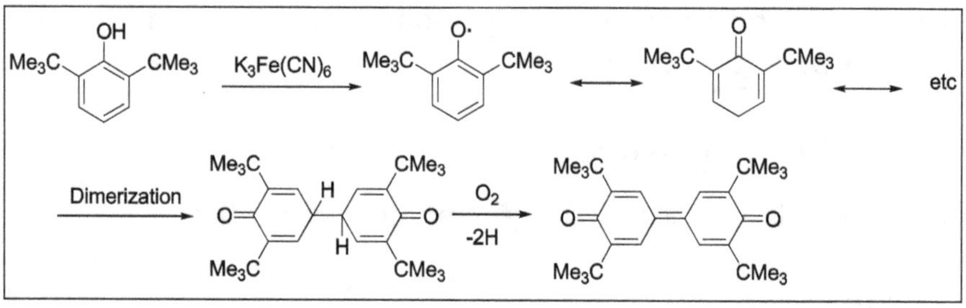

Reactions of Free Radicals

Four types of reactions are possible with free-radicals.

Abstraction

Free radicals proceed reactions with saturated organic molecules by abstracting an atom from carbon. The selectivity of the free radicals towards C-H bonds of different types is determined by bond dissociation energy and polar effects. The rate of the abstraction process increases as bond dissociation energy decreases. For examples:

Bond	$H–CH_3$	$H–CH_2Me$	$H–CHMe_2$	$H–CMe_3$
Bond Dissociation Energy ($KJmol^{-1}$)	426	401	385	372
Reactivity Order	$H–CH_3$ <	HCH_2Me <	$H–CHMe_2$ <	$H–CMe_3$

Allylic and benzylic C-H bonds are weaker than those of saturated systems and exhibit greater reactivity and selectivity with free radicals.

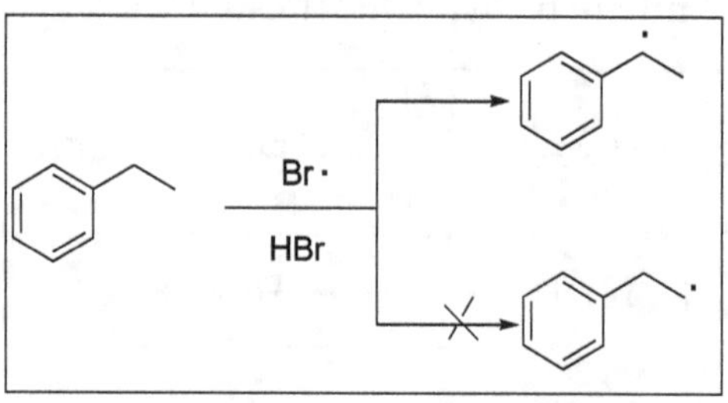

Secondly, polar effect is also operative in radical reactions. For example, the relative reactivity of C-H bonds in butyl chloride towards chlorine atom follows:

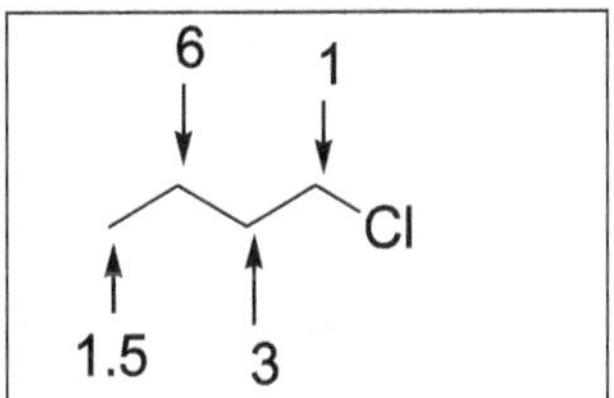

Number refers relative electron density.

Chlorine atom is electronegative and preferentially reacts at C-H bonds of relatively high electron density. In contrast, alkyl radicals which are called nucleophilic radicals react preferentially at CH bonds with electron density low.

Addition

Free radicals undergo addition to carbon-carbon double bonds. The reaction is generally selective. For example, addition to $CH_2 = CHX$ takes place exclusively at the methyl group, irrespective of the nature of X.

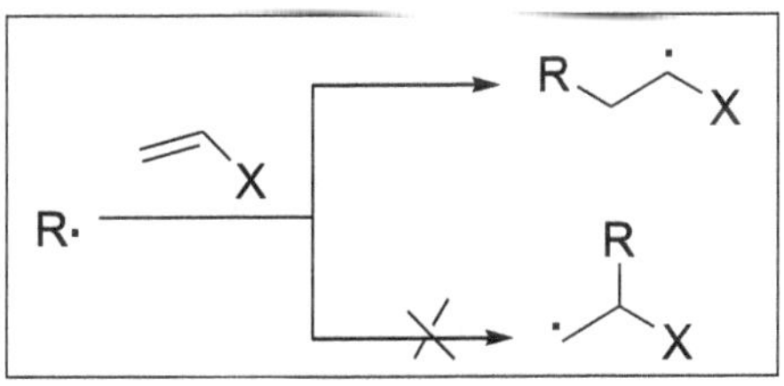

Alkynes also exhibit similar reactivity with free radicals.

Combination and Disproportionation

Two free radicals can combine by dimerization or disproportionation.

$$2Br\cdot \rightarrow Br_2$$
$$2CH_3\cdot \rightarrow H_3C-CH_3$$
$$CH_4\cdot + Br\cdot \rightarrow H_3C-Br$$
$$2Et\cdot \rightarrow EtH + CH_2 = CH_2$$

These reactions are generally fast, some having negligible activation energies.

Rearrangement

Free radicals are known to rearrange in certain circumstances. For example, the addition of free radical leads to rearrangement due to relief of strain in the cyclic system.

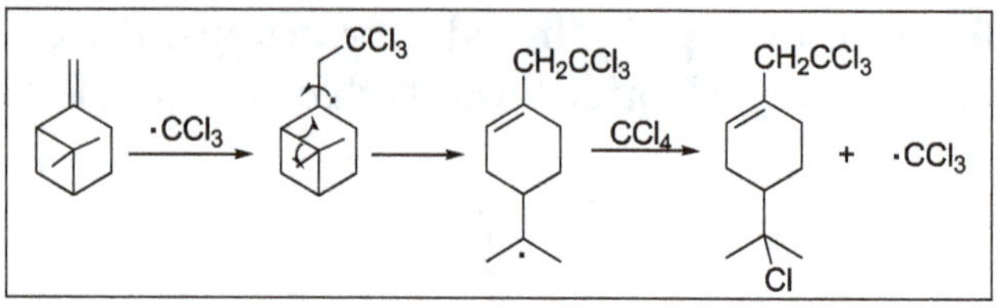

The Characteristics of Free Radical Reactions

Free radical reactions may be broadly classified into two groups. In the first, the product formation occurs from the combination of two radicals.

$$2CH_3 \cdot \rightarrow H_3C-CH_3$$

In the second class, the product formation results from the reaction of a radical with a molecule by chain process.

Initiation	$Cl_2 \xrightarrow{\text{light}} 2Cl\cdot$
Propogation	$Cl \cdot + CH_4 \rightarrow HCl + \cdot CH_3$ $\cdot CH_3 + Cl_2 \rightarrow CH_3Cl + \cdot Cl$
Termination	$2Cl\cdot \rightarrow Cl_2$ $2CH_3\cdot \rightarrow C_2H_6$ $\cdot CH_3 + Cl\cdot \rightarrow CH_3Cl$

General Reaction of Alkanes

Alkane halogenation is an example of a substitution reaction, a type of reaction that often occurs in organic chemistry. A substitution reaction is a chemical reaction in which part of a small reacting molecule replaces an atom or a group of atoms on a hydrocarbon or hydrocarbon derivative.

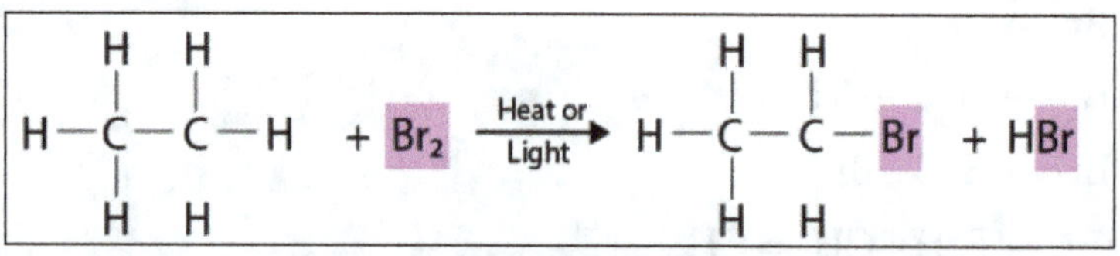

A general equation for the substitution of a single halogen atom for one of the hydrogen atom of an alkane is,

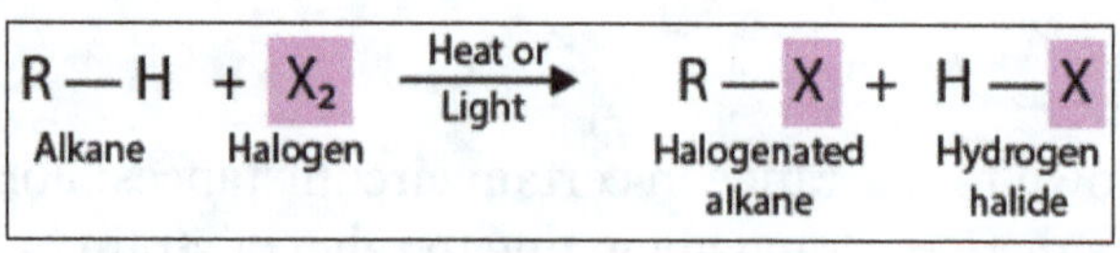

General Features of Halogenation of Alkanes

Following features of halogenation of alkanes:

- The notation R-H is a general formula for an alkane. R in this case represents an alkyl group. Addition of a hydrogen atom to an alkyl group produces the parent hydrocarbon of the alkyl group.
- The notation R-X on the product side is the general formula for a halogenated alkane. X is the general symbol for a halogen atom.
- Reaction conditions are noted by placing these conditions on the equation arrow that separates reactants from products. Halogenation of an alkane requires the presence of heat or light.

Chlorination of Methane by Substitution

In halogenation of an alkane, the alkane is said to undergo fluorination, chlorination, bromination or iodination depending on the identity of the halogen reactant. Chlorination and bromination are the two widely used alkane halogenation reactions. Fluorination reactions generally proceed too quickly to be useful and iodination reactions go too slowly.

Halogenations usually result in the formation of a mixture of products rather than a single product. More than one product results because more than one hydrogen atom on an alkane can be replaced with halogen atoms. Methane and chlorine when heated to a high temperature in the presence of light react as follows:

$$CH_4 \xrightarrow[hv]{Cl_2} \begin{matrix} CH_3Cl \\ + HCl \end{matrix} \xrightarrow[hv]{Cl_2} \begin{matrix} CH_2Cl_2 \\ + HCl \end{matrix} \xrightarrow[hv]{Cl_2} \begin{matrix} CHCl_3 \\ + HCl \end{matrix} \xrightarrow[hv]{Cl_2} \begin{matrix} CCl_4 \\ + HCl \end{matrix}$$

The mechanism for this reaction takes place in three steps:

1. Initiation Step: The Cl-Cl bond of elemental chlorine undergoes hemolysis when irradiated with UV light, and this process yields two chlorine atoms, also called chlorine radicals.

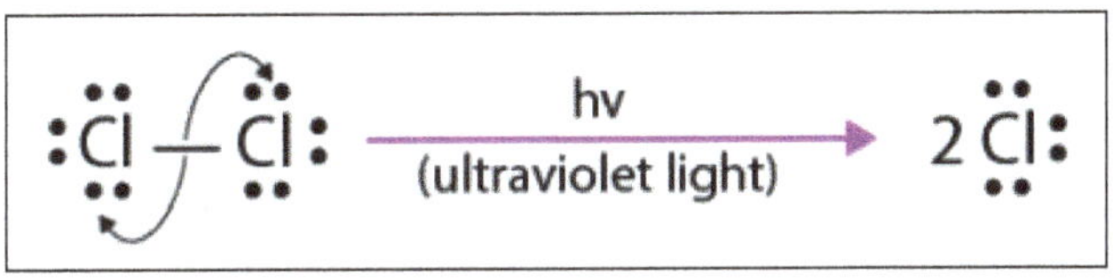

2. Propagation Step: A chlorine radical abstracts a hydrogen atom from methane to produce the methyl radical. The methyl radical in turn abstracts a chlorine atom from a chlorine molecule and chloromethane is formed. The second step of propagation also regenerates a chlorine atom. These steps repeat many times until termination occurs.

$$CH_3Cl + Cl_2 \xrightarrow[\text{light}]{\text{Heat or}} CH_2Cl_2 + HCl$$

$$CH_2Cl_2 + Cl_2 \xrightarrow[\text{light}]{\text{Heat or}} CHCl_3 + HCl$$

$$CHCl_3 + Cl_2 \xrightarrow[\text{light}]{\text{Heat or}} CCl_4 + HCl$$

3. Termination Step: Termination takes place when a chlorine atom reacts with another chlorine atom to generate Cl_2, or chlorine atom can react with a methyl radical to form chloromethane which constitutes a minor pathway by which the product is made. Two methyl radicals can also combine to produce ethane, a very minor by product of this reaction.

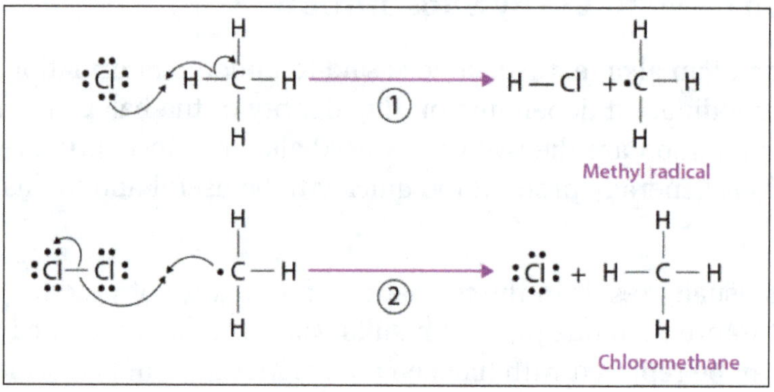

The reaction does not stop at this step, however because the chlorinated methane product can react with additional chlorine to produce polychlorinated products.

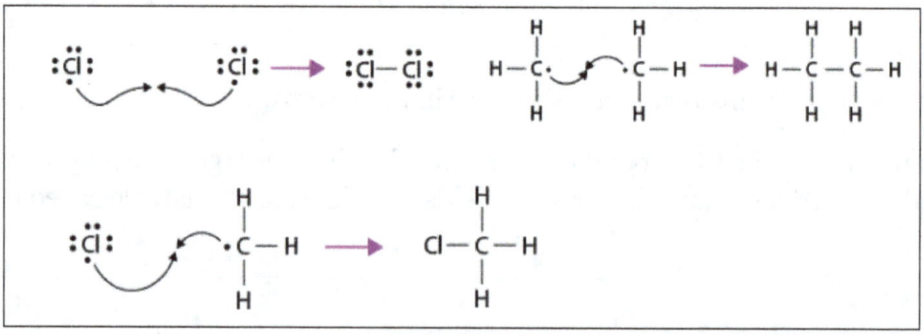

By controlling the reaction conditions and the ratio of chlorine to methane. It is possible to favour formation of one or another of the possible chlorinated methane products.

Electrophilic Substitution Reaction

Benzene reacts with bromine or chlorine in an electrophilic substitution reaction only in the presence of a catalyst which is either chloride or iron. Iron is not a catalyst because it changes permanently during the reaction. It reacts with some bromine to form iron 3 chloride $FeCl_3$ or iron 3 bromide, $FeBr_3$.

$$2Fe + 3Cl_2 \rightarrow 2FeCl_3$$

$$2Fe + 3Br_2 \rightarrow 2FeBr_3$$

Mechanism of Halogenation of Benzene

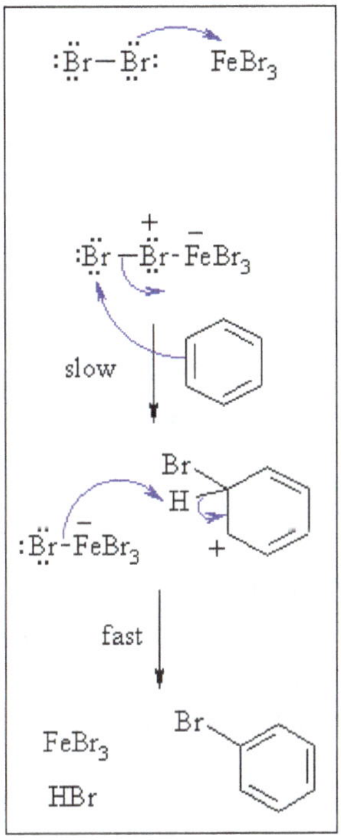

- Step 1: The bromine reacts with Lewis acid to create a complex that makes bromine more electrophilic.
- Step 2: The π electrons of aromatic C = C behave as a nucleophile which attacks the electrophilic Br and displaces iron tetrabromide.
- Step 3: The proton is removed from sp3 C and bears the Bromo group that reforms C=C and aromatic system generates HBr and regulates active catalyst.

The compounds act as a catalyst and behave similarly to aluminium chloride in the following chemical reactions. We will further look at how halogenation is influenced by each halogen or the halogenating agents.

- Reaction with Chlorine: The reaction between chlorine and benzene gives chlorobenzene in the presence of either iron or aluminium chloride.

$$C_6H_6 + Cl_2 \rightarrow C_6H_5Cl + HCl$$

Or,

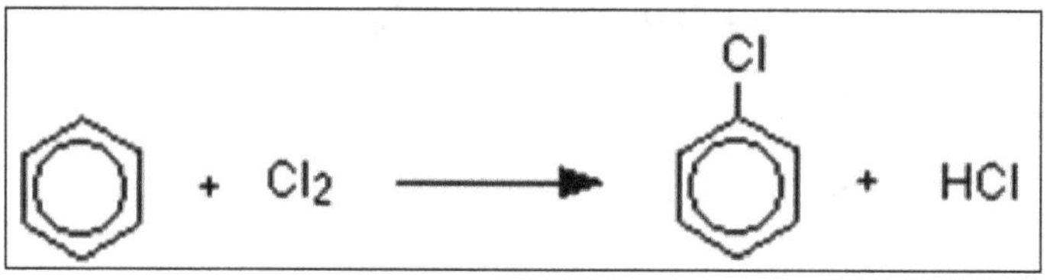

- Reaction with Bromine: The reaction between bromine and benzene gives bromobenzene in the presence of either iron or aluminium bromide. Iron is used normally because it is readily available and cheaper.

  $$C_6H_6 + Br_2 \rightarrow C_6H_5Br + HBr$$

  Or,

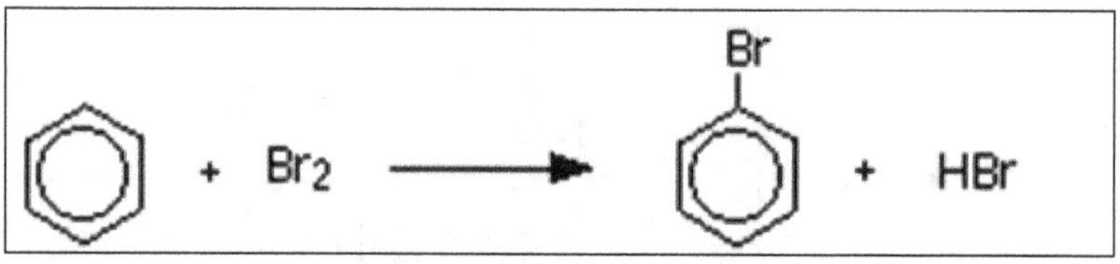

- Reaction with Fluorine: If we take any organic compounds it readily reacts with fluorine (usually explosive in nature). However, in the case of elemental fluorine (F_2) we will need some proper apparatus and fulfil certain conditions. A variety of fluorinating reagents such as xenon difluoride and cobalt (III) fluoride are also used sometimes for the reaction.
- Reaction with Iodine: In comparison to the above three halogens, iodine is the least reactive halogen. It does not readily react with organic substances. Simultaneously, when we study inorganic chemistry, almost every element excluding helium, argon and neon form fluorides when it is reacted with fluorine.

Sulfonation and Sulfation

Sulfonation and sulfation are major industrial chemical processes used to make a diverse range of products, including dyes and color intensifiers, pigments, medicinals, pesticides and organic intermediates.

Although sulfonates and sulfates are similar in structure, there are important differences. Figure shows the reaction to produce a sulfonate. Sulfur trioxide (SO_3) reacts with an organic molecule – in this case an alkyl benzene – to form a sulfur-carbon bond. One of the characteristics of this process is that the resultant alkyl benzene sulfonic acid is a stable molecule.

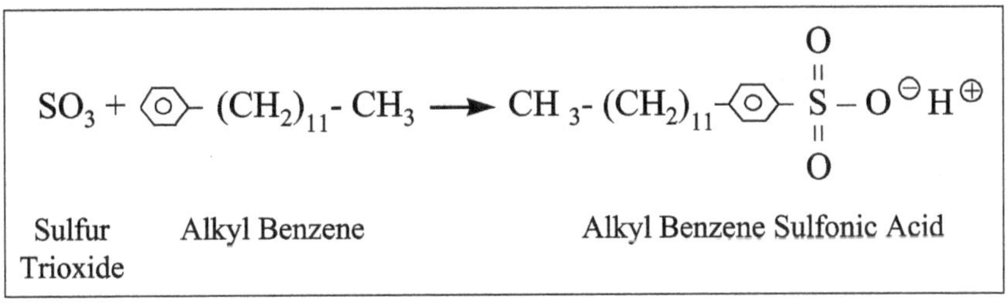

Sulfonation

Sulfation, on the other hand, involves forming a carbon-oxygen-sulfur bond as shown in Figure. The resultant alcohol sulfuric acid is not hydrolytically stable. Unless neutralized, it decomposes to form sulfuric acid and the original alcohol.

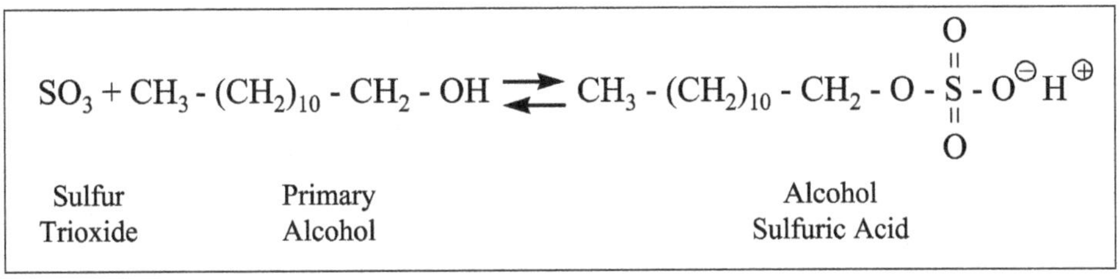

Sulfation

Because they are stable, sulfonic acids can be isolated, stored and shipped as an article of commerce. Sulfates, due to their instability, are available only as neutral compounds. This stability difference in the products of reaction with SO_3 also has a profound impact on the choice of process used to produce sulfonates or sulfates. Some processes, such as oleum sulfonation, cannot be used to make alcohol sulfates containing a low level of inorganic sulfate. However others, such as sulfamic acid sulfation, cannot be used to make sulfonic acids.

SO_3 is an aggressive electrophilic reagent that rapidly reacts with any organic compound containing an electron donor group. Sulfonation is a difficult reaction to perform on an industrial scale because the reaction is rapid and highly exothermic releasing approximately 380 kJ/kg SO_3 (800 BTUs per pound of SO_3) reacted. Most organic compounds form a black char on contact with pure SO_3 due to the rapid reaction and heat evolution. Additionally, as shown in Figure, the reactants increase in viscosity between 15 and 300 times as they are converted from the organic feedstock to the sulfonic acid. This large increase in viscosity makes heat removal difficult. The high viscosity of the formed products reduces the heat transfer coefficient from the reaction mass. Effective cooling of the reaction mass is essential because high temperatures promote side reactions that produce undesirable by-products. Also, precise control of the molar ratio of SO_3 to organic is essential because any excess SO_3, due to its reactive nature, contributes to side reactions and by-product formation. Therefore, commercial scale sulfonation reactions require special equipment and instrumentation that allows tight control of the mole ratio of SO_3 to organic and rapid removal of the heat of reaction.

Table: Viscosity Increase on Sulfonation.

Feedstock	Feed Viscosity (cp)	Acid Viscosity @ 40 - 50 °C
Linear alkyl benzene	5	400
Branched alkyl benzen	15	1000
Ethoxylated alcohol	20	500
Tallow alcohol	10	150
Alpha olefins	3	1000

Historically, the problem of SO_3 reactivity has been solved by diluting and/or complexing the SO_3 to moderate the rate of reaction. Commercially, the diluting or complexing agents include ammonia (sulfamic acid), hydrochloric acid (chlorosulfuric acid), water or sulfuric acid (sulfuric acid or oleum) and dry air (air/SO_3 film sulfonation). Control of the ratio of SO_3 to organic raw material is vital to achieving the desired product quality with use of any of the agents. Additionally, these processes require heat removal to maintain product quality.

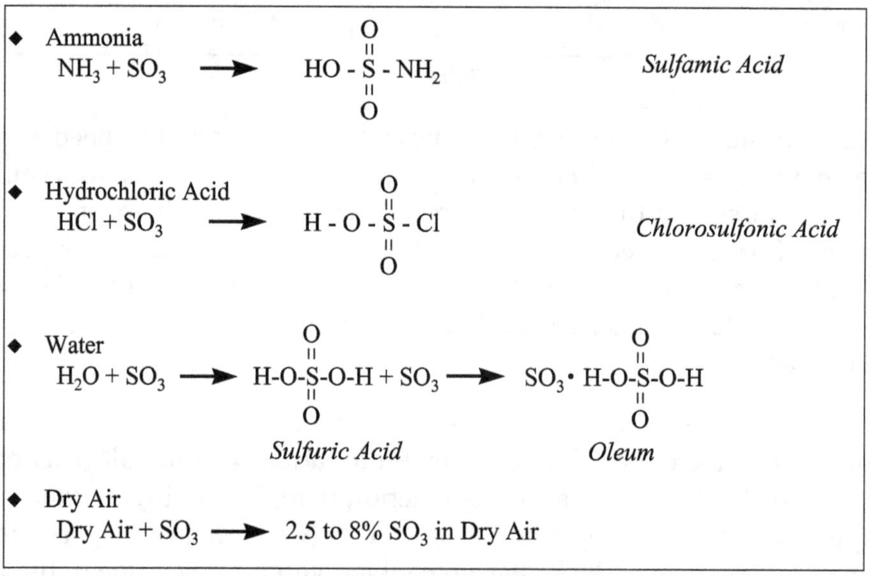

Agents to Reduce SO_3 Reactivity.

Commercial Sulfonation Processes

Sulfamic acid (NH_2SO_3H) is used to sulfate alcohols and ethoxylated alcohols to form an ammonium neutralized salt. A typical reaction is shown in Figure. The reaction goes directly to the ammonium salt of the alcohol sulfuric acid. Sulfamic acid is an expensive reagent, costing approximately US\$0.51 per pound of reactive SO_3. Sulfamic acid sulfation is a mild and specific sulfating reagent suitable for making ammonium neutralized alcohol ethoxylates. Another major advantage of sulfamic acid is that it selectively sulfates alcohol groups and will not sulfonate aromatic rings. Therefore, its major use is sulfation of alkyl phenol ethoxylates. This specificity prevents formation of mixed sulfate-sulfonate compounds. Sulfamic acid is easily handled and reacts stoichiometrically with the alcohol or ethoxy alcohol. It readily adapts to making small quantities of material in low cost batch equipment.

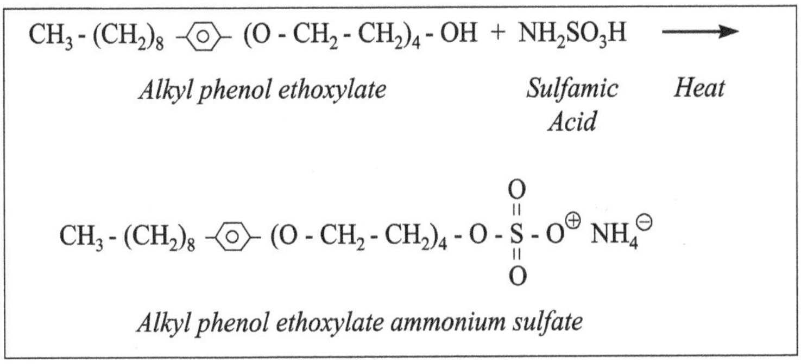

Sulfamic Acid Sulfation.

Chlorosulfuric acid ($ClSO_3H$) is also widely used to produce alcohol sulfates, alcohol ether sulfates, dyes and dye intermediates. Figure shows a typical reaction. As the reaction moves to completion, hydrochloric acid (HCl) is released. This acid must be scrubbed or otherwise recovered. Chlorosulfuric acid is an expensive source of SO_3 although it is approximately one half the cost of sulfamic acid. The cost per pound of reactive SO_3 is US$0.255. It is a rapid, stoichiometric reactant. However, it is still more expensive than other sources of SO_3. It is also corrosive and a hazardous chemical to handle and liberates HCl as a by-product during the reaction. The HCl can be recovered by scrubbing the off-gas stream with water, or neutralized by scrubbing the off-gas with a dilute basic scrubbing solution. In either case, additional equipment and complexity are added to the process.

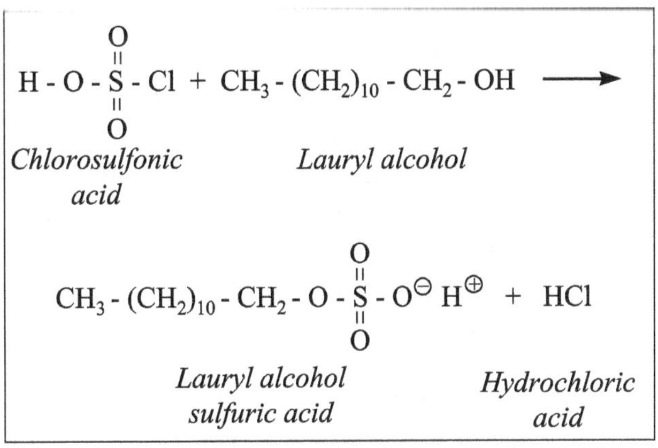

Chlorosulfonic Acid Sulfation.

Sulfuric acid (H_2SO_4) and oleum ($SO_3 \bullet H_2SO_4$) are widely used as sulfonating agents. Oleum is used to sulfonate alkyl benzene and sulfate fatty alcohols for heavy duty detergents. The reaction is shown in Figure. It is an equilibrium process, as water is formed in the reaction and the resultant water dilutes the oleum and/or sulfuric acid. The sulfonation reaction stops when the sulfuric acid concentration drops to approximately

90%. This "spent" acid may be separated from alkyl benzene sulfonic acid to produce a product, which on neutralization contains a relatively low level (6-10%) of sodium sulfate. When fatty alcohols are sulfated, the spent acid cannot be separated. It must be neutralized with the alcohol sulfuric acid to make a product containing a high level of sodium sulfate. Oleum is relatively inexpensive – about US$0.153 per pound of reactive SO_3. Oleum sulfonation can be operated as either a batch or continuous process. This process has the dual advantage of low SO_3 cost and low capital equipment cost. However, it has the disadvantage of being an equilibrium process which leaves large quantities of un-reacted sulfuric acid. This waste acid must be separated from the reaction mixture and subsequently disposed. In North America, the expense of disposing the "spent" sulfuric acid has become so high that the process economics are now questionable.

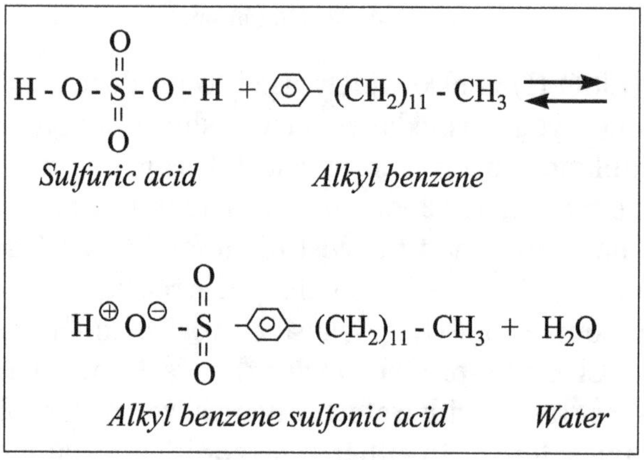

Sulfuric Acid/Oleum.

Sulfonation with sulfuric acid is a special case of oleum sulfonation. Because the sulfonation reaction stops when the acid concentration in the reaction mixture drops to less than approximately 90%, sulfonation of detergent feed-stocks with sulfuric acid is not normally practiced. Today sulfuric acid sulfonation is principally used for production of hydrotropes by azeotropic reaction with benzene, toluene or xylene. In this special process, the water formed during the reaction is removed by azeotropic distillation of the water and unreacted feedstock. The water is then separated from the immiscible organic feedstock which is returned to the reaction vessel. Because water is removed, the reaction may continue to completion.

The air/SO_3 sulfonation process is a direct process in which SO_3 gas is diluted with very dry air and reacted directly with the organic feedstock. The source of the SO_3 gas may be either liquid SO_3 or SO_3 produced by burning sulfur. As shown in Figure, the reaction of gaseous SO_3 with organic is rapid and stoichiometric. It is complicated by the possibility of side reactions and therefore tight process control is essential. The cost for liquid SO_3 is US$0.09 per pound of reactive SO_3; while SO_3 from sulfur burning is US$0.02 per pound of reactive SO_3. The air/SO_3 sulfonation process is the lowest SO_3 cost of any sulfonation process and is extremely versatile, producing very high quality products.

However, it is a continuous process best suited to large production volumes. In addition, it requires expensive precision equipment and highly trained operating personnel.

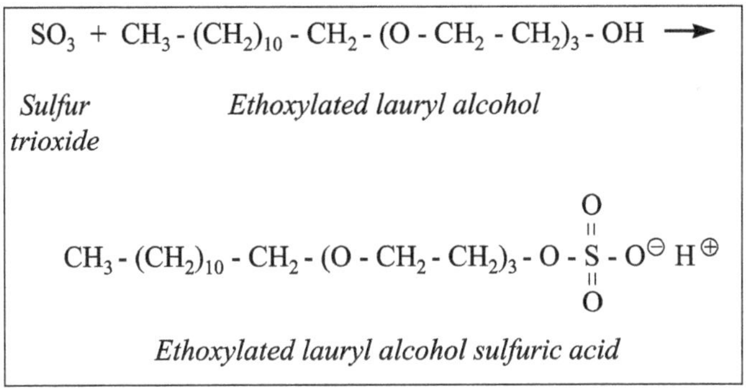

Air/SO_3.

A commercially successful sulfonation process requires reaction of SO_3 with the organic feedstock under tightly controlled conditions. Figure illustrates the level of control demanded by the air/SO_3 sulfonation process. This illustration shows the production of 1, 4-dioxane, an undesirable by-product formed during the sulfation of ethoxylated alcohols. The 1, 4-dioxane formed is a function of mole ratio (kg moles per unit time of SO_3 fed to the reactor divided by kg moles per unit time of feedstock fed to the reactor). As the mole ratio of SO_3 to organic feedstock increases, the level of dioxane in the product remains relatively low at 20 to 30 ppm. A critical point of over-sulfation occurs at a mole ratio of approximately 1.03. Once over-sulfation occurs and the mole ratio exceeds 1.04, dioxane production increases rapidly to values measured in hundreds of parts per million. Similar adverse responses are observed with product color or levels of un-sulfonated or un-sulfated (free oil) materials in the product. Clearly, the sulfonation process must be controlled to within 1% of the desire mole ratio in order to achieve excellent product quality. Other important process variables are reaction temperature, SO_3 gas concentration, time to neutralization, neutralization pH and neutralization temperature. These variables also influence product quality although the effect is not as dramatic as the effect of mole ratio.

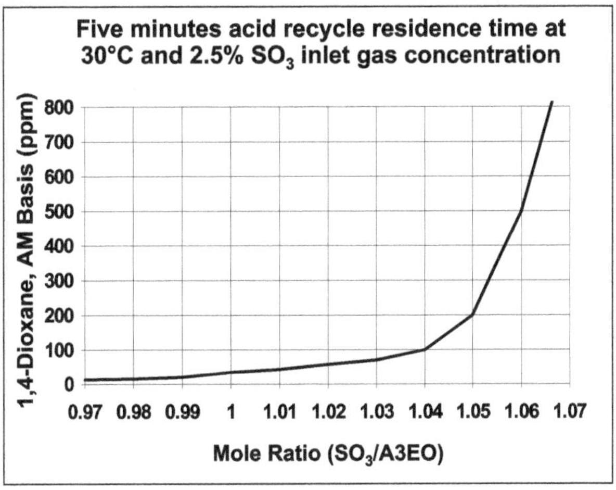

1,4-Dioxane vs. Mole Ratio.

References

- Apostila-bmassa: feq.unicamp.br, Retrieved 25, March 2020
- Mechanism-of-nitration-1: chemvista.org, Retrieved 11, July 2020
- Mechanism-of-nitration-2: chemvista.org, Retrieved 04, May 2020
- Halogenation-of-alkanes, chemisty: byjus.com, Retrieved 21, August 2020
- Sulfo-and-Sulfa, Technical-papers-201: chemithon.com, Retrieved 08, February 2020

3

Methods of Separation

The process to convert a mixture of chemical substances into two or more distinct product mixtures is termed as chemical separation. A few techniques used for chemical separation are chromatography, extraction, filtration and centrifugation. All these chemical separation techniques have been carefully analyzed in this chapter.

Separation Process

Separation techniques are those techniques that can be used to separate two different states of matter such as liquid and solid. Such separation techniques include filtration or evaporation. Separation process, or a separation method, or simply a separation, is a methodology to attain any mass transfer phenomenon that converts a mixture of substances into two or more distinct product mixtures.

Separations based on Size

Size is the simplest physical property we can exploit in a separation. To accomplish the separation we use a porous medium through which only the analyte or the interferent can pass. Examples of size-based separations include filtration, dialysis, and size-exclusion. In a filtration we separate a particulate interferent from dissolved analytes using a filter whose pore size retains the interferent. The solution passing through the filter is called the filtrate, and the material retained by the filter is the retentate. Gravity filtration and suction filtration using filter paper are techniques with which you should already be familiar. Membrane filters, available in a variety of micrometer pores sizes, are the method of choice for particulates that are too small to be retained by filter paper. Figure provides information about three types of membrane filters.

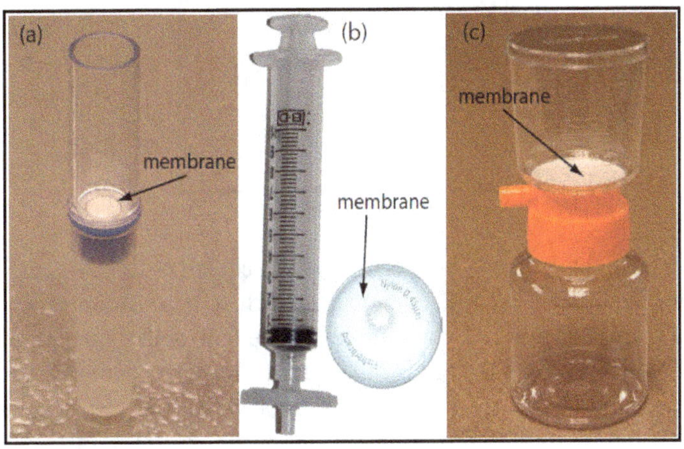

In the figure above, three types of membrane filters for separating analytes from interferents. (a) A centrifugal filter for concentrating and desalting macromolecular solutions. The membrane has a nominal molecular weight cut-off of 1 × 106 g/mol. The sample is placed in the upper reservoir and the unit is placed in a centrifuge. Upon spinning the unit at 2000×g–5000×g, the filtrate collects in the bottom reservoir and the retentate remains in the upper reservoir. (b) A 0.45 μm membrane syringe filter. The photo on the right shows the membrane filter in its casing. In the photo on the left, the filter is attached to a syringe. Samples are placed in the syringe and pushed through the filter. The filtrate is collected in a test tube or other suitable container. (c) A disposable filter system with a 0.22 μm cellulose acetate membrane filter. The sample is added to the upper unit and vacuum suction is used to draw the filtrate through the membrane and into the lower unit. To store the filtrate, the top half of the unit is removed and a cap placed on the lower unit. The filter unit shown here has a capacity of 150 mL.

Dialysis is another example of a separation technique that uses size to separate the analyte and the interferent. A dialysis membrane is usually constructed from cellulose and fashioned into tubing, bags, or cassettes. Figure shows an example of a commercially available dialysis cassette. The sample is injected into the dialysis membrane, which is sealed tightly by a gasket, and the unit is placed in a container filled with a solution whose composition is different from the sample. If the concentration of a particular species is different on the membrane's two sides, the resulting concentration gradient provides a driving force for its diffusion across the membrane. While small species freely pass through the membrane, larger species are unable to pass. Dialysis is frequently used to purify proteins, hormones, and enzymes. During kidney dialysis, metabolic waste products, such as urea, uric acid, and creatinine, are removed from blood by passing it over a dialysis membrane.

Figure: Example of a dialysis cassette.

The dialysis membrane in this unit has a molecular weight cut-off of 10 000 g/mol. Two sheets of the membrane are separated by a gasket and held in place by the plastic frame. Four ports, one of which is labeled, provide a means for injecting the sample between the dialysis membranes. The cassette is inverted and submerged in a beaker containing the external solution, which is stirred using a stir bar. A foam buoy, used as a stand in the photo, serves as a float so that the unit remains suspended above the stir bar. The external solution is usually replaced several time during dialysis. When dialysis is complete, the solution remaining in the cassette is removed through an injection port.

Size-exclusion chromatography is a third example of a separation technique that uses size as a means for effecting a separation. In this technique a column is packed with small, approximately 10-μm, porous polymer beads of cross-linked dextrin or polyacrylamide. The pore size of the particles is controlled by the degree of cross-linking, with more cross-linking producing smaller pore sizes. The sample is placed into a stream of solvent that is pumped through the column at a fixed flow rate. Those species too large to enter the pores pass through the column at the same rate as the solvent. Species entering into the pores take longer to pass through the column, with smaller species needing more time to pass through the column. Size-exclusion chromatography is widely used in the analysis of polymers, and in biochemistry, where it is used for the separation of proteins.

Separations based on Mass or Density

If the analyte and the interferent differ in mass or density, then a separation using centrifugation may be possible. The sample is placed in a centrifuge tube and spun at a high angular velocity, measured in revolutions per minute (rpm). The sample's constituents experience a centrifugal force that pulls them toward the bottom of the centrifuge tube. The species experiencing the greatest centrifugal force has the fastest sedimentation rate, and is the first to reach the bottom of the centrifuge tube. If two species have equal density, their separation is based on mass, with the heavier species having the greater sedimentation rate. If the species are of equal mass, then the species with the largest density has the greatest sedimentation rate.

Centrifugation is an important separation technique in biochemistry. For example, lists conditions for separating selected cellular components. We can separate lysosomes from other cellular components by several differential centrifugations, in which we divide the sample into a solid residue and a supernatant solution. After destroying the cells, the solution is centrifuged at 15 000 × g (a centrifugal force that is 15 000 times that of the earth's gravitational force) for 20 minutes, leaving a solid residue of cell membranes and mitochondria. The supernatant, which contains the lysosomes, is isolated by decanting it from the residue and centrifuged at 30 000 × g for 30 minutes, leaving a solid residue of lysosomes. Figure shows a typical centrifuge capable of produce the centrifugal forces needed for biochemical separations.

Table: Conditions for Separating Selected Cellular Components by Centrifugation

Components	Centrifugal Force (×g)	Time (min)
eukaryotic cells	1000	5
cell membranes, nuclei	4000	10
mitochondria, bacterial cells	15000	20
lysosomes, bacterial membranes	30000	30
ribosomes	100000	180

Figure: *Bench-top centrifuge capable of reaching speeds up to 14 000 rpm and centrifugal forces of 20 800 × g. This particular centrifuge is refrigerated, allowing samples to be cooled to temperatures as low as –4° C.*

An alternative approach to differential centrifugation is density gradient centrifugation. To prepare a sucrose density gradient, for example, a solution with a smaller concentration of sucrose—and, thus, of lower density—is gently layered upon a solution with a higher concentration of sucrose. Repeating this process several times, fills the centrifuge tube with a multi-layer density gradient. The sample is placed on top of the density gradient and centrifuged using a force greater than 150 000 × *g*. During centrifugation, each of the sample's components moves through the gradient until it reaches a position where its density matches the surrounding sucrose solution. Each component is isolated as a separate band positioned where its density is equal to that of the local density within the gradient. Figure below provides an example of a typical sucrose density centrifugation for separating plant thylakoid membranes.

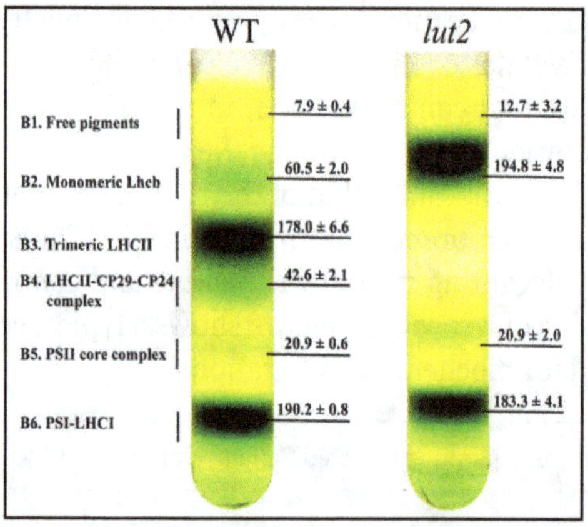

Figure: Example of a sucrose density gradient centrifugation of thylakoid membranes from wild type (WT) and lut2 plants. The thylakoid membranes were extracted from the plant's leaves and separated by centrifuging in a 0.1–1 M sucrose gradient for 22 h at 280 000 × g at 4° C. Six bands and their chlorophyll contents are shown.

Separations based on Complexation Reactions

One widely used technique for preventing an interference is to bind the interferent in a strong, soluble complex that prevents it from interfering in the analyte's determination. This process is known as masking. As shown in Table, a wide variety of ions and molecules are useful masking agents, and, as a result, selectivity is usually not a problem.

Table: Selected Inorganic and Organic Masking Agents for Metal Ions

Masking Agent	Elements Whose Ions Can Be Masked
CN^-	Ag, Au, Cd, Co, Cu, Fe, Hg, Mn, Ni, Pd, Pt, Zn
SCN^-	Ag, Cd, Co, Cu, Fe, Ni, Pd, Pt, Zn
NH_3	Ag, Co, Ni, Cu, Zn
F^-	Al, Co, Cr, Mg, Mn, Sn, Zn
$S_2O_3^{2-}$	Au, Ce, Co, Cu, Fe, Hg, Mn, Pb, Pd, Pt, Sb, Sn, Zn
tartrate	Al, Ba, Bi, Ca, Ce, Co, Cr, Cu, Fe, Hg, Mn, Pb, Pd, Pt, Sb, Sn, Zn
oxalate	Al, Fe, Mg, Mn
thioglycolic acid	Cu, Fe, Sn

Separations based on a Change of State

Because an analyte and its interferent are usually in the same phase, we can achieve a separation if one of them undergoes a change in its physical state or its chemical state.

Changes in Physical State

When the analyte and the interferent are miscible liquids, a distillation may be possible if their boiling points are significantly different. Figure below shows the progress of a distillation as a plot of temperature versus the mixture's vapor-phase and liquid-phase composition. The initial mixture (point A), contains more interferent than analyte. When this solution is brought to its boiling point, the vapor phase in equilibrium with the liquid phase is enriched in analyte (point B). The horizontal line connecting points A and B represents this vaporization equilibrium. Condensing the vapor phase at point B, by lowering the temperature, creates a new liquid phase whose composition is identical to that of the vapor phase (point C). The vertical line connecting points B and C represents this condensation equilibrium. The liquid phase at point C has a lower boiling point than the original mixture, and is in equilibrium with the vapor phase at point D. This process of repeated vaporization and condensation gradually separates the analyte and interferent.

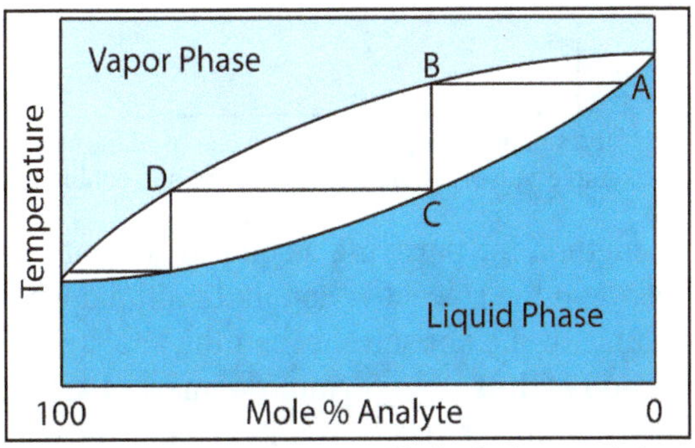

Figure: Boiling point versus composition diagram for a near-ideal solution consisting of a low-boiling analyte and a high-boiling interferent. The horizontal lines represent vaporization equilibria and the vertical lines represent condensation equilibria.

Two experimental set-ups for distillations are shown in figure. The simple distillation apparatus shown in Figure (a) does not produce a very efficient separation, and is useful only for separating a volatile analyte (or interferent) from a non-volatile interferent (or analyte), or for an analyte and interferent whose boiling points differ by more than 150° C. A more efficient separation is achieved by a fractional distillation. Packing the distillation column with a high surface area material, such as a steel sponge or glass beads, provides more opportunity for the repeated process of vaporization and condensation necessary to effect a complete separation.

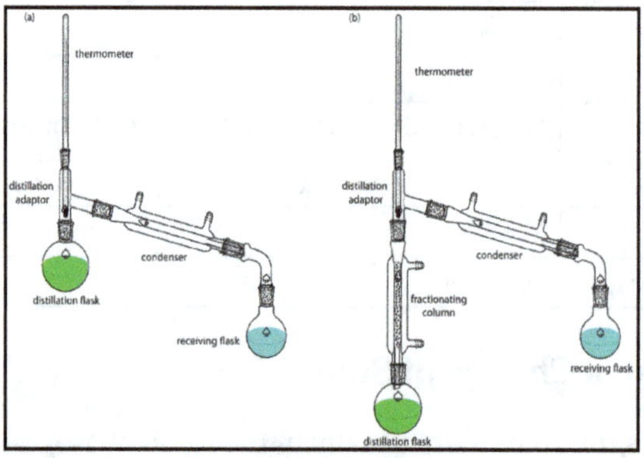

Figure: Typical experimental set-up for (a) a simple distillation, and (b) a fractional distillation.

When the sample is a solid, sublimation may provide a useful separation of the analyte and the interferent. The sample is heated at a temperature and pressure below the analyte's triple point, allowing it to vaporize without passing through the liquid state. Condensing the vapor recovers the purified analyte. A useful analytical example of sublimation is the isolation of amino acids from fossil mollusk shells and deep-sea sediments.

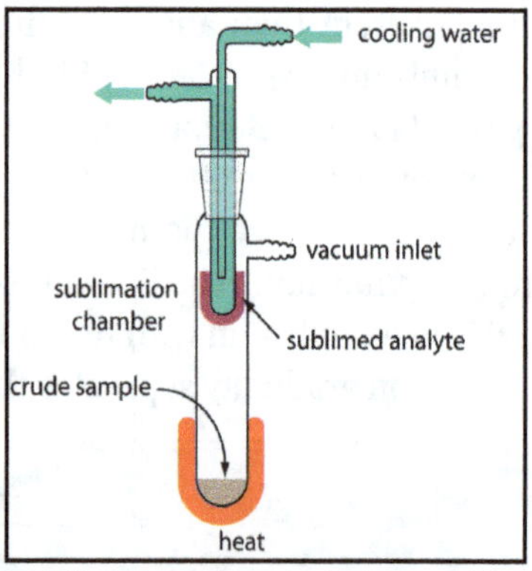

Figure: Typical experimental setup for a sublimation. The sample is placed in the sublimation chamber, which may be evacuated. Heating the sample causes the analyte to vaporize and sublime onto a cooled probe.

Recrystallization is another method for purifying a solid. A solvent is chosen in which the analyte's solubility is significant when the solvent is hot, and minimal when the solvent is cold. The interferents must be less soluble in the hot solvent than the analyte, or present in much smaller amounts. After heating a portion of the solvent in an Erlenmeyer flask, small amounts of the sample are added until undissolved sample is visible. Additional hot solvent is added until the sample redissolves, or until only insoluble impurities remain. This process of adding sample and solvent is repeated until the entire sample has been added to the Erlenmeyer flask. Any insoluble impurities are removed by filtering the hot solution. The solution is allowed to cool slowly, promoting the growth of large, pure crystals, and then cooled in an ice bath to minimize solubility losses. The purified sample is isolated by filtration and rinsed to remove any soluble impurities. Finally, the sample is dried to remove any remaining traces of the solvent. Further purification, if necessary, can be accomplished by additional recrystallizations.

Changes in Chemical State

Distillation, sublimation, and recrystallization use a change in physical state as a means of separation. Chemical reactivity also can be a useful tool for separating analytes and interferents. For example, we can separate SiO_2 from a sample by reacting it with HF to form SiF_4. Because SiF_4 is volatile, it is easy to remove by evaporation. If we wish to collect the reaction's volatile product, then a distillation is possible. For example, we can isolate the NH_4^+ in a sample by making the solution basic and converting it to NH_3. The ammonia is then removed by distillation. Table provides additional examples of this approach for isolating inorganic ions.

Table: Examples of Using a Chemical Reaction and a Distillation to Separate an Inorganic Analyte From Interferents

Analyte	Treatment	Isolated Species
CO_3^{2-}	$CO_3^{2-}(aq) + 2H_3O^+(aq) \rightarrow CO_2(g) + 3H_2O(l)$	CO_2
NH_4^+	$NH_4^+(aq) + OH^-(aq) \rightarrow NH_3(g) + H_2O(l)$	NH_3
SO_3^{2-}	$SO_3^{2-}(aq) + 2H_3O^+(aq) \rightarrow SO_2(g) + 3H_2O(l)$	SO_2
S^{2-}	$S^{2-}(aq) + 2H_3O+(aq) \rightarrow H_2S(g) + 2H_2O(l)$	H_2S

Another reaction for separating analytes and interferents is precipitation. Two important examples of using a precipitation reaction in a separation are the pH-dependent solubility of metal oxides and hydroxides, and the solubility of metal sulfides.

Separations based on the pH-dependent solubility of oxides and hydroxides usually use strong acids, strong bases, or an NH_3/NH_4Cl buffer. Most metal oxides and hydroxides are soluble in hot concentrated HNO_3, although a few oxides, such as WO_3, SiO_2, and SnO_2 remain insoluble even under these harsh conditions. In determining the amount of Cu in brass, for example, we can avoid an interference from Sn by dissolving the sample with a strong acid and filtering to remove the solid residue of SnO_2.

Most metals form a hydroxide precipitate in the presence of concentrated NaOH. Those metals forming amphoteric hydroxides, however, do not precipitate because they react to form higher-order hydroxo-complexes. For example, Zn^{2+} and Al^{3+} do not precipitate in concentrated NaOH because the form the soluble complexes $Zn(OH)_3^-$ and $Al(OH)_4^-$. The solubility of Al^{3+} in concentrated NaOH allows us to isolate aluminum from impure samples of bauxite, an ore of Al_2O_3. After crushing the ore, we place it in a solution of concentrated NaOH, dissolving the Al_2O_3 and forming $Al(OH)_4^-$. Other oxides in the ore, such as Fe_2O_3 and SiO_2, remain insoluble. After filtering, we recover the aluminum as a precipitate of $Al(OH)_3$ by neutralizing some of the OH^- with acid.

The pH of an NH_3/NH_4Cl buffer ($pK_a = 9.26$) is sufficient to precipitate most metals as the hydroxide. The alkaline earths and alkaline metals, however, do not precipitate at this pH. In addition, metal ions that form soluble complexes with NH_3, such as Cu^{2+}, Zn^{2+}, Ni^{2+}, and Co^{2+} also do not precipitate under these conditions.

The use of S^{2-} as a precipitating reagent is one of the earliest examples of a separation technique. In Fresenius's 1881 text *A System of Instruction in Quantitative Chemical Analysis*, sulfide is frequently used to separate metal ions from the remainder of the sample's matrix.[14] Sulfide is a useful reagent for separating metal ions for two reasons:

(1) Most metal ions, except for the alkaline earths and alkaline metals, form insoluble sulfides;

(2) These metal sulfides show a substantial variation in solubility. Because the concentration of S^{2-} is pH-dependent, we can control which metal precipitates by adjusting the pH. For example, in Fresenius's gravimetric procedure for determining Ni in ore samples, sulfide is used three times as a means of separating Co^{2+} and Ni^{2+} from Cu^{2+} and, to a lesser extent, from Pb^{2+}.

Separations Based on a Partitioning Between Phases

The most important group of separation techniques uses a selective partitioning of the analyte or interferent between two immiscible phases. If we bring a phase containing a solute, S, into contact with a second phase, the solute partitions itself between the two phases, as shown by the following equilibrium reaction.

$$S_{\text{phase 1}} \rightleftharpoons S_{\text{phase 2}}$$

The equilibrium constant for reaction $S_{\text{phase 1}} \rightleftharpoons S_{\text{phase 2}}$

$$K_D = \frac{\left[S_{phase\,2}\right]}{\left[S_{phase\,1}\right]}$$

is called the distribution constant or partition coefficient. If K_D is sufficiently large, then the solute moves from phase 1 to phase 2. The solute remains in phase 1 if the partition coefficient is sufficiently small. When we bring a phase containing two solutes into contact with a second phase, if K_D is favorable for only one of the solutes a separation of the solutes is possible. The physical states of the phases are identified when describing the separation process, with the phase containing the sample listed first. For example, if the sample is in a liquid phase and the second phase is a solid, then the separation involves liquid–solid partitioning.

Extraction between Two Phases

We call the process of moving a species from one phase to another phase an extraction. Simple extractions are particularly useful for separations where only one component has a favorable partition coefficient. Several important separation techniques are based on a simple extraction, including liquid–liquid, liquid–solid, solid–liquid, and gas–solid extractions.

Liquid–Liquid Extractions

A liquid–liquid extraction is usually accomplished with a separatory funnel. After placing the two liquids in the separatory funnel, we shake the funnel to increase the surface area between the phases. When the extraction is complete, we allow the liquids to separate. The stopcock at the bottom of the separatory funnel allows us to remove the two phases.

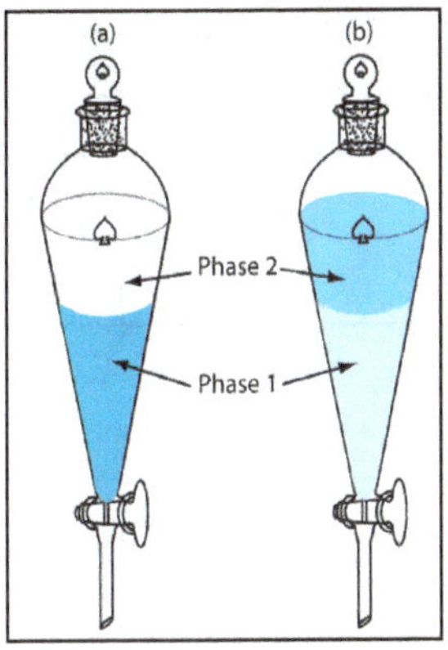

Figure: Example of a liquid–liquid extraction using a separatory funnel. (a) Before the extraction, 100% of the analyte is in phase 1. (b) After the extraction, most of the analyte is in phase 2, although some analyte remains in phase 1. Although one liquid–liquid extraction can result in the complete transfer of analyte, a single extraction usually is not sufficient.

We also can carry out a liquid–liquid extraction without a separatory funnel by adding the extracting solvent to the sample container. Pesticides in water, for example, are preserved in the field by extracting them into a small volume of hexane. Liquid–liquid microextractions, in which the extracting phase is a 1-mL drop suspended from a microsyringe.

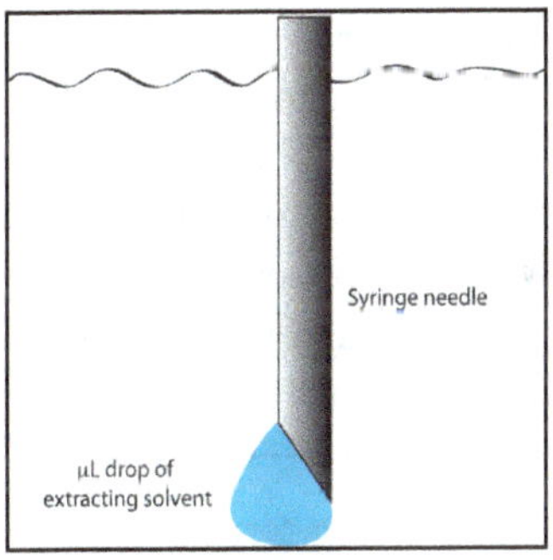

Figure: *Schematic of a liquid–liquid* microextraction showing a syringe needle with a μL drop of the extracting solvent.

Solid Phase Extractions

In a solid phase extraction of a liquid sample, we pass the sample through a cartridge containing a solid adsorbent, several examples of which are shown in figure. The choice of adsorbent is determined by the species we wish to separate. Table provides several representative examples of solid adsorbents and their applications.

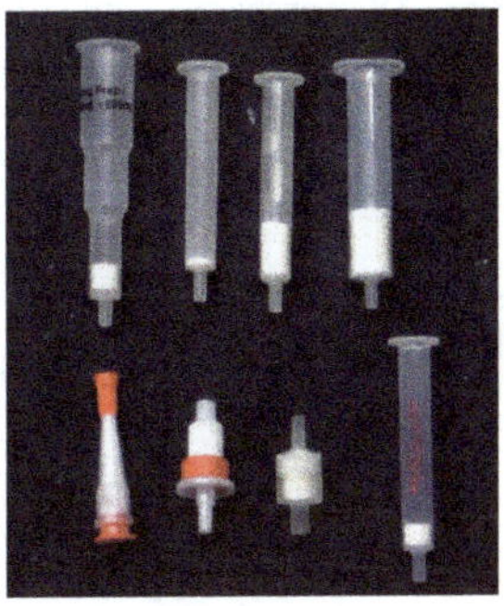

Table: Representative Adsorbents for the Solid Phase Extraction of Liquid Samples.

Adsorbent	Structure	Properties and Uses
silica	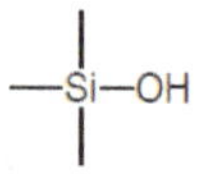	• retains low to moderate polarity species from organic matricies • fat soluble vitamins, steroids
aminopropyl	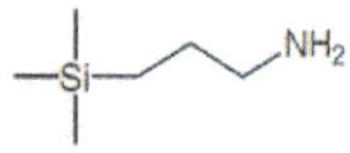	• retains polar compounds • carbohydrates, organic acids
cyanopropyl	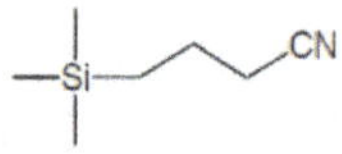	• retains wide variety of species from aqueous and organic matricies • pesticides, hydrophobic peptides
diol	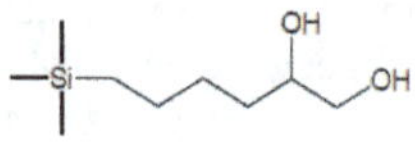	• retains wide variety of species from aqueous and organic matricies • proteins, peptides, fungicides
octadecyl (C-18)	$-C_{18}H_{37}$	• retains hydrophobic species from aqueous matricies • caffeine, sedatives, polyaromatic hydrocarbons, carbohydrates, pesticides
octyl (C-8)	$-C_8H_{17}$	• similar to C-18

As an example, let's examine a procedure for isolating the sedatives secobarbital and phenobarbital from serum samples using a C-18 solid adsorbent. Before adding the sample, the solid phase cartridge is rinsed with 6 mL each of methanol and water. Next, a 500-μL sample of serum is pulled through the cartridge, with the sedatives and matrix interferents retained by a liquid–solid extraction. Washing the cartridge with distilled water removes any interferents. Finally, we elute the sedatives using 500 μL of acetone. In comparison to a liquid–liquid extraction, a solid phase extraction has the advantage of being easier, faster, and requiring less solvent.

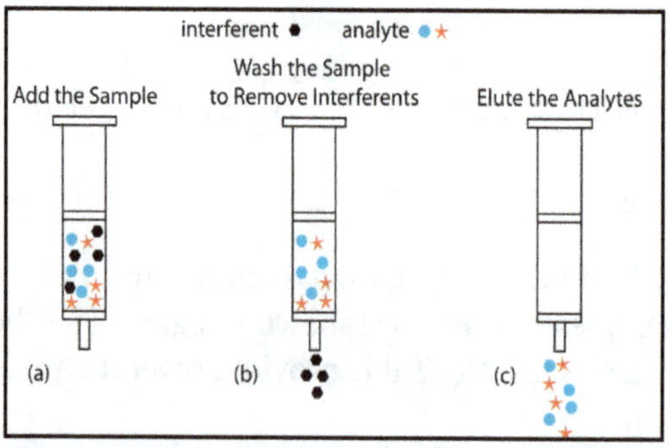

Figure: Steps in a typical solid phase extraction. After preconditioning the solid phase cartridge with solvent, (a) the sample is added to the cartridge, (b) the sample is washed to remove interferents, and (c) the analytes eluted.

Continuous Extractions

An extraction is possible even if the analyte has an unfavorable partition coefficient, provided that the sample's other components have significantly smaller partition coefficients. Because the analyte's partition coefficient is unfavorable, a single extraction will not recover all the analyte. Instead we continuously pass the extracting phase through the sample until a quantitative extraction is achieved.

A continuous extraction of a solid sample is carried out using a Soxhlet extractor. The extracting solvent is placed in the lower reservoir and heated to its boiling point. Solvent in the vapor phase moves upwards through the tube on the far right side of the apparatus, reaching the condenser where it condenses back to the liquid state. The solvent then passes through the sample, which is held in a porous cellulose filter thimble, collecting in the upper reservoir. When the solvent in the upper reservoir reaches the return tube's upper bend, the solvent and extracted analyte are siphoned back to the lower reservoir. Over time the analyte's concentration in the lower reservoir increases.

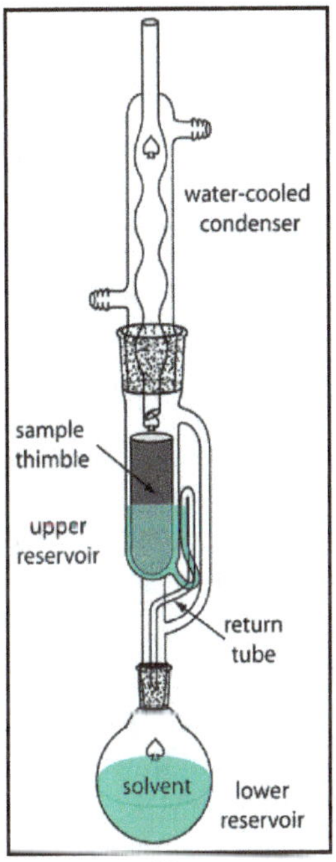

Figure: Soxhlet extractor.

Microwave-assisted extractions have replaced Soxhlet extractions in some applications. The process is the same as that described earlier for a microwave digestion. After placing the sample and the solvent in a sealed digestion vessel, a microwave oven is used to heat the mixture. Using a sealed digestion vessel allows the extraction to take place at a higher temperature and pressure, reducing the amount of time needed for a quantitative extraction. In a Soxhlet extraction the temperature is limited by the solvent's boiling point at atmospheric pressure. When acetone is the solvent, for example, a Soxhlet extraction is limited to 56° C, but a microwave extraction can reach 150° C.

Two other continuous extractions deserve mention. Volatile organic compounds (VOCs) can be quantitatively removed from liquid samples by a liquid–gas extraction. As shown in figure, an inert purging gas, such as He, is passed through the sample. The purge gas removes the VOCs, which are swept to a primary trap where they collect on a solid absorbent. A second trap provides a means for checking to see if the primary trap's capacity is exceeded. When the extraction is complete, the VOCs are removed from the primary trap by rapidly heating the tube while flushing with He. This technique is known as a purge-and-trap. Because the analyte's recovery may not be reproducible, an internal standard is necessary for quantitative work.

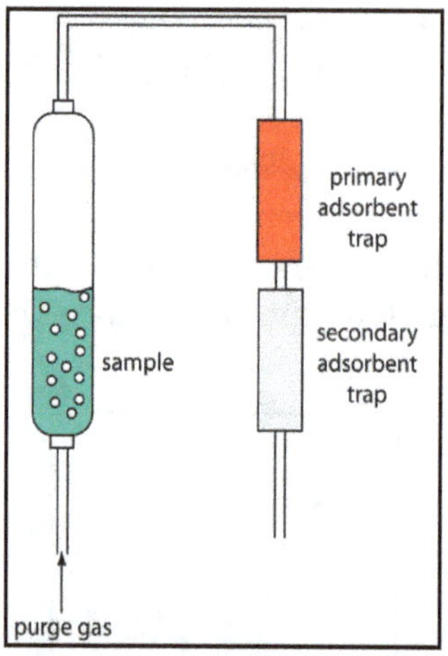

Figure: Schematic diagram of a purge-and-trap system for extracting volatile analytes. The purge gas releases the analytes, which subsequently collect in the primary adsorbent trap. The secondary adsorption trap is monitored for evidence of the analyte's breakthrough.

Continuous extractions also can be accomplished using supercritical fluids. If we heat a substance above its critical temperature and pressure, it forms a supercritical fluid whose properties are between those of a gas and a liquid. A supercritical fluid is a better solvent than a gas, which makes it a better reagent for extractions. In addition, a supercritical fluid's viscosity is significantly less than that of a liquid, making it easier to push it through a particulate sample. One example of a supercritical fluid extraction is the determination of total petroleum hydrocarbons (TPHs) in soils, sediments, and sludges using supercritical CO_2. An approximately 3-g sample is placed in a 10-mL stainless steel cartridge and supercritical CO_2 at a pressure of 340 atm and a temperature of 80°C is passed through the cartridge for 30 minutes at flow rate of 1–2 mL/min. To collect the TPHs, the effluent from the cartridge is passed through 3 mL of tetrachloroethylene at room temperature. At this temperature the CO_2 reverts to the gas phase and is released to the atmosphere.

Chromatographic Separations

In an extraction, the sample is one phase and we extract the analyte or the interferent into a second phase. We also can separate the analyte and interferents by continuously passing one sample-free phase, called the mobile phase, over a second sample-free phase that remains fixed or stationary. The sample is injected into the mobile phase and the sample's components partition themselves between the mobile phase and the stationary phase. Those components with larger partition coefficients are more likely to move into the stationary phase, taking a longer time to pass through the system. This is the basis of all chromatographic separations. Chromatography provides both a separation of analytes and interferents, and a means for performing a qualitative or quantitative analysis for the analyte.

Chromatography as a Separation Process

Most of us have got our papers wet at some time or another, but have you ever noticed what happens to the ink as the water spreads? It doesn't always smudge and blur, as you might expect.

Sometimes it splits up into weird colored streaks that creep across the page. When that happens, you're seeing chromatography in action. In this case it's totally accidental, but we can also use it by design to split up mixtures and other substances into their components. Chromatography is actually one of the most useful analytical techniques chemists have at their disposal, helpful in everything from identifying biological materials to finding clues at crime scenes.

Chromatography is based on the principle where molecules in mixture applied onto the surface or into the solid, and fluid stationary phase (stable phase) is separating from each other while moving with the aid of a mobile phase. The factors effective on this separation process include molecular characteristics related to adsorption (liquid-solid), partition (liquid-solid), and affinity or differences among their molecular weights. Because of these differences, some components of the mixture stay longer in the stationary phase, and they move slowly in the chromatography system, while others pass rapidly into mobile phase, and leave the system faster.

Based on this approach three components form the basis of the chromatography technique.

- Stationary phase: This phase is always composed of a "solid" phase or "a layer of a liquid adsorbed on the surface a solid support".
- Mobile phase: This phase is always composed of "liquid" or a "gaseous component."
- Separated molecules.

The type of interaction between stationary phase, mobile phase, and substances contained in the mixture is the basic component effective on separation of molecules from each other. Chromatography methods based on partition are very effective on separation, and identification of small molecules as amino acids, carbohydrates, and fatty acids. However, affinity chromatographies (ie. ion-exchange chromatography) are more effective in the separation of macromolecules as nucleic acids, and proteins. Paper chromatography is used in the separation of proteins, and in studies related to protein synthesis; gas-liquid chromatography is utilized in the separation of alcohol, esther, lipid, and amino groups, and observation of enzymatic interactions, while molecular-sieve chromatography is employed especially for the determination of molecular weights of proteins. Agarose-gel chromatography is used for the purification of RNA, DNA particles, and viruses.

Stationary phase in chromatography, is a solid phase or a liquid phase coated on the surface of a solid phase. Mobile phase flowing over the stationary phase is a gaseous or liquid phase. If mobile phase is liquid it is termed as liquid chromatography (LC), and if it is gas then it is called gas chromatography (GC). Gas chromatography is applied for gases, and mixtures of volatile liquids, and solid material. Liquid chromatography is used especially for thermal unstable, and non-volatile samples.

The purpose of applying chromatography which is used as a method of quantitative analysis apart from its separation, is to achive a satisfactory separation within a suitable timeinterval. Various chromatography methods have been developed to that end. Some of them include column chromatography, thin-layer chromatography (TLC), paper chromatography, gas chromatography, ion exchange chromatography, gel permeation chromatography, high-pressure liquid chromatography, and affinity chromatography.

Column Chromatography

Since proteins have difference characteristic features as size, shape, net charge, stationary phase used, and binding capacity, each one of these characteristic components can be purified using chromatographic methods. Among these methods, most frequently column chromatography is applied. This technique is used for the purification of biomolecules. On a column (stationary phase) firstly the sample to be separated, then wash buffer (mobile phase) are applied. Their flow through inside column material placed on a fiberglass support is ensured. The samples are accumulated at the bottom of the device in a time-dependent and volume-dependent manner.

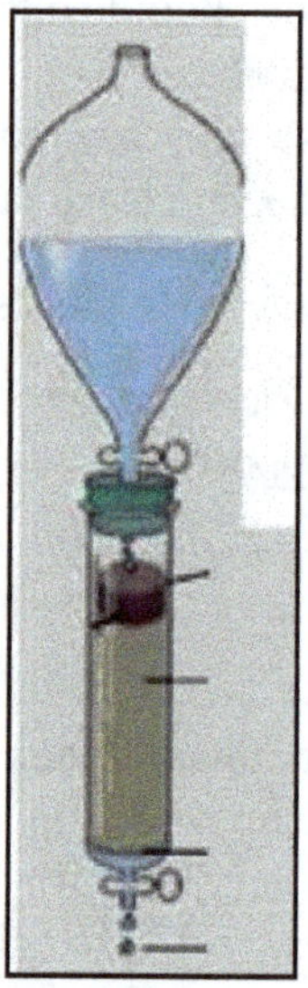

Figure: Column chromatography.

Ion-exchange Chromatography

Ion-exchange chromatography is based on electrostatic interactions between charged protein groups, and solid support material (matrix). Matrix has an ion load opposite to that of the protein to be separated, and the affinity of the protein to the column is achieved with ionic ties. Proteins are separated from the column either by changing pH, concentration of ion salts or ionic strength of the buffer solution. Positively charged ion-exchange matrices are called anion-exchange matrices, and adsorb negatively charged proteins. While matrices bound with negatively charged groups are known as cation-exchange matrices, and adsorb positively charged proteins.

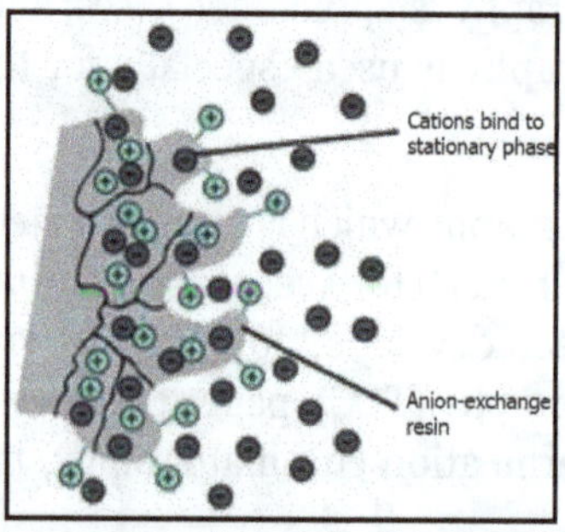

Figure: Ion-exchange chromatography.

Gel- permeation (Molecular Sieve) Chromatography

The basic principle of this method is to use dextran containing materials to separate macromolecules based on their differences in molecular sizes. This procedure is basically used to determine molecular weights of proteins, and to decrease salt concentrations of protein solutions. In gel-permeation column stationary phase consists of inert molecules with small pores. The solution containing molecules of different dimensions are passed continuously with a constant flow rate through the column. Molecules larger than pores cannot permeate into gel particles, and they are retained between particles within a restricted area. Larger molecules pass through spaces between porous particles, and move rapidly through inside the column. Molecules smaller than the pores are diffused into pores, and as molecules get smaller, they leave the column with proportionally longer retention times. Sephadeks G type is the most frequently used column material. Besides, dextran, agorose, polyacrylamide are also used as column materials.

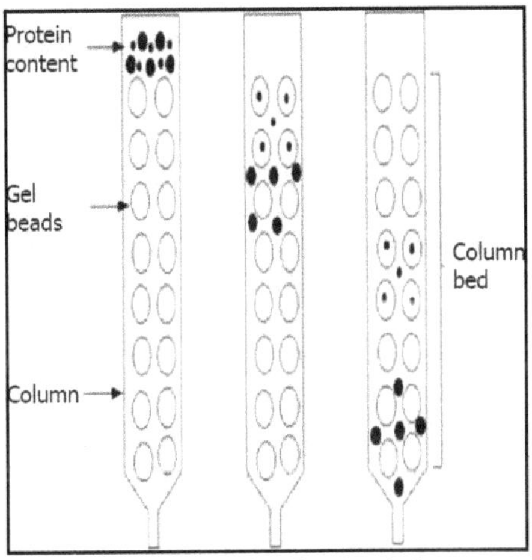

Figure: Gel-permeation (molecular sieve) chromatography.

Affinity Chromatography

Figure: Affinity chromatography.

This chromatography technique is used for the purification of enzymes, hormones, antibodies, nucleic acids, and specific proteins. A ligand which can make a complex with specific protein (dextran, polyacrylamide, cellulose etc.) binds the filling material of the column. The specific protein which makes a complex with the ligand is attached to the solid support (matrix), and retained in the column, while free proteins leave the column. Then the bound protein leaves the column by means of changing its ionic strength through alteration of pH or addition of a salt solution.

Paper Chromatography

In paper chromatography support material consists of a layer of cellulose highly saturated with water. In this method a thick filter paper comprised the support, and water drops settled in its pores made up the stationary "liquid phase." Mobile phase consists of an appropriate fluid placed in a developing tank. Paper chromatography is a "liquid-liquid" chromatography.

Thin-layer Chromatography

Thin-layer chromatography is a "solid-liquid adsorption" chromatography. In this method stationary phase is a solid adsorbent substance coated on glass plates. As adsorbent material all solid substances used. in column chromatography (alumina, silica gel, cellulose) can be utilized. In this method, the mobile phase travels upward through the stationary phase. The solvent travels up the thin plate soaked with the solvent by means of capillary action. During this procedure, it also drives the mixture priorly dropped on the lower parts of the plate with a pipette upwards with different flow rates. Thus the separation of analytes is achieved. This upward travelling rate depends on the polarity of the material, solid phase, and of the solvent.

In cases where molecules of the sample are colorless, florescence, radioactivity or a specific chemical substance can be used to produce a visible colored reactive product so as to identify their positions on the chromatogram. Formation of a visible color can be observed under room light or UV light. The position of each molecule in the mixture can be measured by calculating the ratio between the distances travelled by the molecule and the solvent. This measurement value is called relative mobility, and expressed with a symbol R_f. R_f value is used for qualitative description of the molecules.

Gas Chromatography

In this method stationary phase is a column which is placed in the device, and contains a liquid stationary phase which is adsorbed onto the surface of an inert solid. Gas chromatography is a "gas-liquid" chromatography. Its carrier phase consists of gases as He or N_2. Mobile phase which is an inert gas is passed through a column under high pressure. The sample to be analyzed is vaporized, and enters into a gaseous mobile phase phase. The components contained in the sample are dispersed between mobile phase and stationary phase on the solid support. Gas chromatography is a simple, multifaceted, highly sensitive, and rapidly applied technique for the extremely excellent separation of very minute molecules. It is used in the separation of very little amounts of analytes.

Dye-ligand Chromatography

Development of this technique was based on the demonstration of the ability of many enzymes to bind purine nucleotides for Cibacron Blue F3GA dye. The planar ring structure with negatively charged groups is analogous to the structure of NAD. This analogy has been evidenced by demonstration of the binding of Cibacron Blue F3GA dye to adenine, ribose binding sites of NAD. The dye behaves as an analogue of ADP-ribose. The binding capacity of this type adsorbents is 10–20-fold stronger than that of the affinity of other adsorbents. Under appropriate pH conditions, elution with high-ionic strength solutions and using ion-exchange property of adsorbent, the adsorbed proteins are separated from the column.

Hydrophobic Interaction Chromatography

In this method the adsorbents prepared as column material for the ligand binding in affinity chromatography are used. HIC technique is based on hydrophobic interactions between side chains bound to chromatography matrix.

Pseudoaffinity Chromatography

Some compounds as anthraquinone dyes, and azo-dyes can be used as ligands because of their affinity especially for dehydrogenases, kinases, transferases, and reeducates The mostly known type of this kind of chromatography is immobilized metal affinity chromatography (IMAC).

High-pressure Liquid Chromatography

Using this chromatography technique it is possible to perform structural, and functional analysis, and purification of many molecules within a short time, This technique yields perfect results in the separation, and identification of amino acids, carbohydrates, lipids, nucleic acids, proteins, steroids, and other biologically active molecules, In HPLC, mobile phase passes through columns under 10–400 atmospheric pressure, and with a high (0.1–5 cm//sec) flow rate. In this technique, use of small particles, and application of high pressure on the rate of solvent flow increases separation power, of HPLC and the analysis is completed within a short time.

Essential components of a HPLC device are solvent depot, high- pressure pump, commercially prepared column, detector, and recorder. Duration of separation is controlled with the aid of a computerized system, and material is accrued.

Applications of Chromatography

Chromatography is one of the important analytical techniques which play an important role in industries especially in pharmaceutical industries and chemical industry.

In Pharmaceutical sector, the main role is to identify and analyse samples for the presence of trace elements or chemical. Its major role is the separation of compounds based on their molecular weight and element composition. Chromatography also detects the unknown compounds and purity of mixture. Chromatography importance is growing its importance in drug development.

In Chemical industry, Chromatography is applied in testing water samples and also checks air quality. HPLC and GC are very much used for detecting various contaminants such as polychlorinated biphenyl (PCBs) in pesticides and oils.

Chromatography also used in various life sciences applications, such as:

- Food Industry: Chromatography has been playing a vital role in Food spoilage and additive detection. In addition it also helps in determining nutritional quality of food.

- Forensic Science: Gas Chromatography has been widely used in forensic pathology and crime scene testing like analysing blood and hair samples of crime place.
- Molecular Biology Studies: Various hyphenated techniques in chromatography such as EC-LC-MS are applied in study of metabolomics and proteomics along with nucleic acids research.

The other applications of chromatography especially HPLC is used in Protein Separation like Insulin Purification, Plasma Fractionation and Enzyme Purification and also in various departments like Fuel Industry, biotechnology and biochemical processes.

Gas Chromatography

Gas chromatography is a term used to describe the group of analytical separation techniques used to analyze volatile substances in the gas phase. In gas chromatography, the components of a sample are dissolved in a solvent and vaporized in order to separate the analytes by distributing the sample between two phases: a stationary phase and a mobile phase. The mobile phase is a chemically inert gas that serves to carry the molecules of the analyte through the heated column. Gas chromatography is one of the sole forms of chromatography that does not utilize the mobile phase for interacting with the analyte. The stationary phase is either a solid adsorbant, termed gas-solid chromatography (GSC), or a liquid on an inert support, termed gas-liquid chromatography (GLC).

In early 1900s, Gas chromatography (GC) was discovered by Mikhail Semenovich Tsvett as a separation technique to separate compounds. In organic chemistry, liquid-solid column chromatography is often used to separate organic compounds in solution. Among the various types of gas chromatography, gas-liquid chromatography is the method most commonly used to separate organic compounds. The combination of gas chromatography and mass spectrometry is an invaluable tool in the identification of molecules. A typical gas chromatograph consists of an injection port, a column, carrier gas flow control equipment, ovens and heaters for maintaining temperatures of the injection port and the column, an integrator chart recorder and a detector.

To separate the compounds in gas-liquid chromatography, a solution sample that contains organic compounds of interest is injected into the sample port where it will be vaporized. The vaporized samples that are injected are then carried by an inert gas, which is often used by helium or nitrogen. This inert gas goes through a glass column packed with silica that is coated with a liquid. Materials that are less soluble in the liquid will increase the result faster than the material with greater solubility.

In GLC, the liquid stationary phase is adsorbed onto a solid inert packing or immobilized on the capillary tubing walls. The column is considered packed if the glass or metal column tubing is packed with small spherical inert supports. The liquid phase adsorbs onto the surface of these beads in a thin layer. In a capillary column, the tubing walls are coated with the stationary phase or an adsorbant layer, which is capable of supporting the liquid phase. However, the method of GSC, has limited application in the laboratory and is rarely used due to severe peak tailing and the semi-permanent retention of polar compounds within the column. Therefore, the method of gas-liquid chromatography is simply shortened to gas chromatography and will be referred to as such here.

Instrumentation

Sample Injection

A sample port is necessary for introducing the sample at the head of the column. Modern injection techniques often employ the use of heated sample ports through which the sample can be injected and vaporized in a near simultaneous fashion. A calibrated microsyringe is used to deliver a sample volume in the range of a few microliters through a rubber septum and into the vaporization chamber. Most separations require only a small fraction of the initial sample volume and a sample splitter is used to direct excess sample to waste. Commercial gas chromatographs often allow for both split and splitless injections when alternating between packed columns and capillary columns. The vaporization chamber is typically heated 50 °C above the lowest boiling point of the sample and subsequently mixed with the carrier gas to transport the sample into the column.

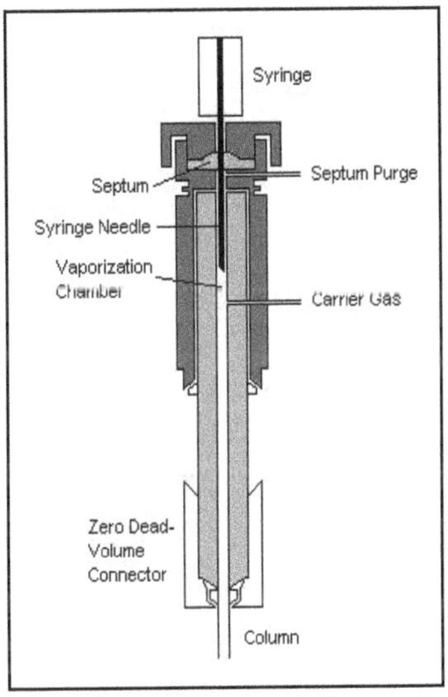

Figure: A cross-sectional view of a microflash vaporizer direct injector.

Carrier Gas

The carrier gas plays an important role, and varies in the GC used. Carrier gas must be dry, free of oxygen and chemically inert mobile-phase employed in gas chromatography. Helium is most commonly used because it is safer than, but comprable to hydrogen in efficiency, has a larger range of flow rates and is compatible with many detectors. Nitrogen, argon, and hydrogen are also used depending upon the desired performance and the detector being used. Both hydrogen and helium, which are commonly used on most traditional detectors such as Flame Ionization(FID), thermal conductivity (TCD) and Electron capture (ECD), provide a shorter analysis time and lower elution temperatures of the sample due to higher flow rates and low molecular weight. For instance, hydrogen or helium as the carrier gas gives the highest sensitivity with TCD because the difference in thermal conductivity between the organic vapor and hydrogen/helium is greater than other carrier gas. Other detectors such as mass spectroscopy, uses nitrogen or argon which has a much better advantage than hydrogen or helium due to their higher molecular weights, in which improve vacuum pump efficiency.

All carrier gasses are available in pressurized tanks and pressure regulators, gages and flow meters are used to meticulously control the flow rate of the gas. Most gas supplies used should fall between 99.995% - 99.9995% purity range and contain a low levels (< 0.5 ppm) of oxygen and total hydrocarbons in the tank. The carrier gas system contains a molecular sieve to remove water and other impurities. Traps are another option to keep the system pure and optimum sensitive and removal traces of water and other contaminants. A two stage pressure regulation is required to use to minimize the pressure surges and to monitor the flow rate of the gas. To monitor the flow rate of the gas a flow or pressure regulator was also require onto both tank and chromatograph gas inlet. This applies different gas type will use different type of regulator. The carrier gas is preheated and filtered with a molecular sieve to remove impurities and water prior to being introduced to the vaporization chamber. A carrier gas is typically required in GC system to flow through the injector and push the gaseous components of the sample onto the GC column, which leads to the detector.

Gas Recommendations for Capillary Columns

Detector	Carrier gas	Preferred makeup gas	Second choice	Detector, anode purge, or reference gas
Electron Capture	Hydrogen	Argon/Methane	Nitrogen	Anode purge must be same as makeup
	Helium	Argon/Methane	Nitrogen	
	Nitrogen	Nitrogen	Argon/Methane	
	Argon/Methane	Argon/Methane	Nitrogen	
Flame Ionization	Hydrogen	Nitrogen	Helium	Hydrogen and air for detector
	Helium	Nitrogen	Helium	
	Nitrogen	Nitrogen	Helium	
Flame Photometric	Hydrogen	Nitrogen		Hydrogen and air for detector
	Helium	Nitrogen		
	Nitrogen	Nitrogen		
	Argon	Nitrogen		
Nitrogen-Phosphorus	Helium	Nitrogen	Helium**	Hydrogen and air for detector
	Nitrogen	Nitrogen	Helium**	
Thermal Conductivity	Hydrogen*	Must be same as carrier and reference gas	Must be same as carrier and reference gas	Reference must be same as carrier and makeup
	Helium			
	Nitrogen			

* When using hydrogen with a thermal conductivity detector, vent the detector exhaust to a fume hood or a dedicated exhaust to avoid buildup of hydrogen gas.

**Helium is not recommended as a makeup gas at flow rates > 5 mL/min. Flow rates above 5 mL/min shorten detector life.

Figure: Gas Recommendations for Capillary Columns.

Gas Recommendations for Packed Columns

Detector	Carrier gas	Comments	Detector, anode purge, or reference gas
Electron Capture	Nitrogen	Maximum sensitivity	Nitrogen
	Argon/Methane	Maximum dynamic range	Argon/Methane
Flame Ionization	Nitrogen	Maximum sensitivity	Hydrogen and air for detector
	Helium	Acceptable alternative	
Flame Photometric	Hydrogen		Hydrogen and air for detector
	Helium		
	Nitrogen		
	Argon		
Nitrogen-Phosphorus	Helium	Optimum performance	Hydrogen and air for detector
	Nitrogen	Acceptable alternative	
Thermal Conductivity	Helium	General use	Reference must be same as carrier
	Hydrogen	Maximum sensitivity (Note A)	
	Nitrogen	Hydrogen detection (Note B)	
	Argon	Maximum hydrogen sensitivity (Note B)	

Note A: Slightly greater sensitivity than helium. Incompatible with some compounds.
Note B: For analysis of hydrogen or helium. Greatly reduces sensitivity for other compounds.

Figure: Gas Recommendations for Packed Columns.

Column Oven

The thermostatted oven serves to control the temperature of the column within a few tenths of a degree to conduct precise work. The oven can be operated in two manners: isothermal programming or temperature programming. In isothermal programming, the temperature of the column is held constant throughout the entire separation. The optimum column temperature for isothermal operation is about the middle point of the boiling range of the sample. However, isothermal programming works best only if the boiling point range of the sample is narrow. If a low isothermal column temperature is used with a wide boiling point range, the low boiling fractions are well resolved but the high boiling fractions are slow to elute with extensive band broadening. If the temperature is increased closer to the boiling points of the higher boiling components, the higher boiling components elute as sharp peaks but the lower boiling components elute so quickly there is no separation.

In the temperature programming method, the column temperature is either increased continuously or in steps as the separation progresses. This method is well suited to separating a mixture with a broad boiling point range. The analysis begins at a low temperature to resolve the low boiling components and increases during the separation to resolve the less volatile, high boiling components of the sample. Rates of 5-7 °C/minute are typical for temperature programming separations.

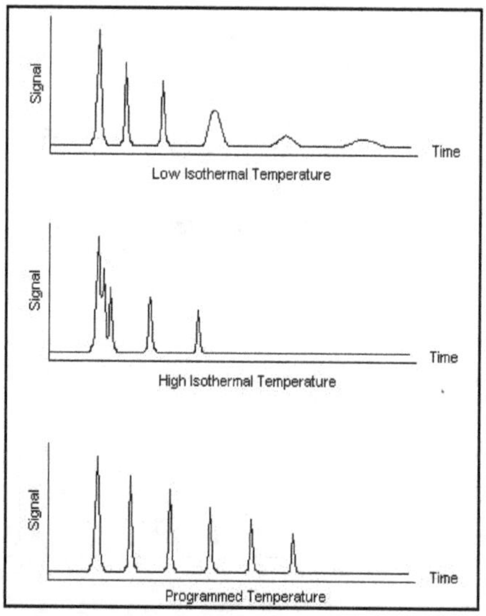

Figure: The effect of column temperature on the shape of the peaks.

Open Tubular Columns and Packed Columns

Open tubular columns, which are also known as capillary columns, come in two basic forms. The first is a wall-coated open tubular (WCOT) column and the second type is a support-coated open tubular (SCOT) column. WCOT columns are capillary tubes that have a thin layer of the stationary phase coated along the column walls. In SCOT columns, the column walls are first coated with a thin layer (about 30 micrometers thick) of adsorbant solid, such as diatomaceous earth, a material which consists of single-celled, sea-plant skeletons. The adsorbant solid is then treated with the liquid stationary phase. While SCOT columns are capable of holding a greater volume of stationary phase than a WCOT column due to its greater sample capacity, WCOT columns still have greater column efficiencies.

Most modern WCOT columns are made of glass, but T316 stainless steel, aluminum, copper and plastics have also been used. Each material has its own relative merits depending upon the application. Glass WCOT columns have the distinct advantage of chemical etching, which is usually achieved by gaseous or concentrated hydrochloric acid treatment. The etching process gives the glass a rough surface and allows the bonded stationary phase to adhere more tightly to the column surface.

One of the most popular types of capillary columns is a special WCOT column called the fused-silica wall-coated (FSWC) open tubular column. The walls of the fused-silica columns are drawn from purified silica containing minimal metal oxides. These columns are much thinner than glass columns, with diameters as small as 0.1 mm and lengths as long as 100 m. To protect the column, a polyimide coating is applied to the outside of the tubing and bent into coils to fit inside the thermostatted oven of the gas chromatography unit. The FSWC columns are commercially available and currently replacing older columns due to increased chemical inertness, greater column efficiency and smaller sampling size requirements. It is possible to achieve up to 400,000 theoretical plates with a 100 m WCOT column, yet the world record for the largest number of theoretical plates is over 2 million plates for 1.3 km section of column.

Packed columns are made of a glass or metal tubing which is densely packed with a solid support like diatomaceous earth. Due to the difficulty of packing the tubing uniformly, these types of columns have a larger diameter than open tubular columns and have a limited range of length. As a result, packed columns can only achieve about 50% of the efficiency of a comparable WCOT column. Furthermore, the diatomaceous earth packing is deactivated over time due to the semi-permanent adsorption of impurities within the column. In contrast, FSWC open tubular columns are manufactured to be virtually free of these adsorption problems.

	Type of Column			
	FSWC	WCOT	SCOT	Packed
Length	10 to 1000 m	10 to 1000 m	10 to 100 m	1 to 6 m
Inner Diameter	0.1 to 0.3 mm	0.25 to 0.75 mm	0.5 mm	2 to 4 mm
Efficiency (plates/m)	2000 to 4000	1000 to 4000	600 to 1200	500 to 1000
Sample Size	10 to 75 ng	10 to 1000 ng	10 to 1000 ng	10 to 10^6 ng
Pressure	Low	Low	Low	High
Speed	Fast	Fast	Fast	Slow
Inertness	Best	Good	Fair	Poor

Figure: Properties of gas chromatography columns.

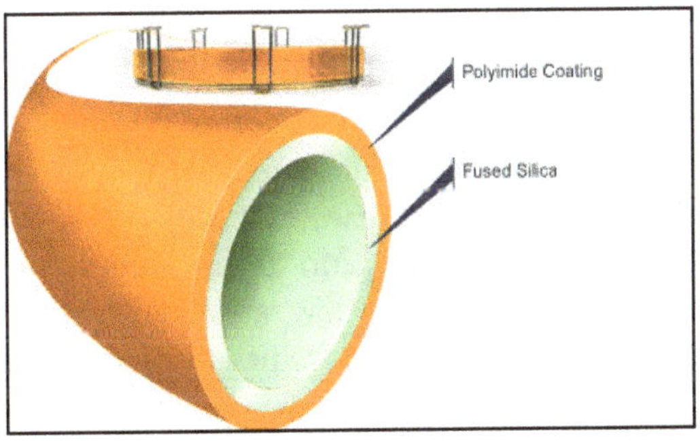

Figure: Computer Generated Image of a FSWC column (specialized for measuring BAC levels).

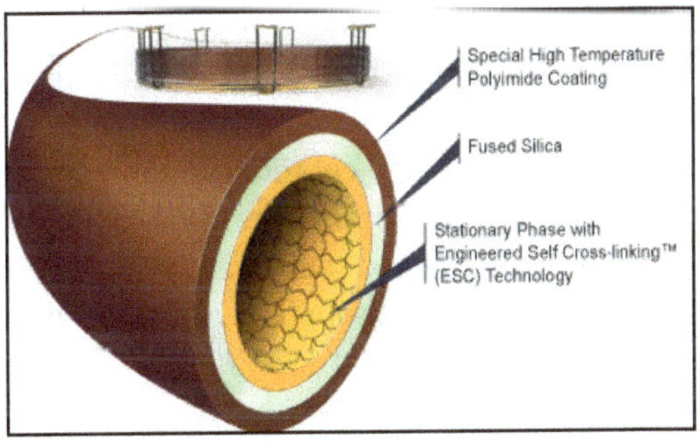

Figure: Computer Generated Image of a FSWC column (specialized to withstand extreme heat).

Different types of columns can be applied for different fields. Depending on the type of sample, some GC columns are better than the others. For example, the FSWC column shown in Figure 5 is designed especially for blood alcohol analysis. It produces fast run times with baseline resolution of key components in less than 3 minutes. Moreover, it displays enhanced resolutions of ethanol and acetone peaks, which helps with determining the BAC levels. This particular column is known as Zebron-BAC and it made with polyimide coating on the outside and the inner layer is made of fused silica and the inner diameter ranges from .18 mm to .25 mm. There are also many other Zebron brand columns designed for other purposes.

Another example of a Zebron GC column is known as the Zebron-inferno. Its outer layer is coated with a special type of polyimide that is designed to withstand high temperatures. As shown in figure, it contains an extra layer inside. It can withstand up to 430 °C to be exact and it is designed to provide true boiling point separation of hydrocarbons distillation methods. Moreover, it is also used for acidic and basic samples.

Detection Systems

The detector is the device located at the end of the column which provides a quantitative measurement of the components of the mixture as they elute in combination with the carrier gas. In theory, any property of the gaseous mixture that is different from the carrier gas can be used as a detection method. These detection properties fall into two categories: bulk properties and specific properties. Bulk properties, which are also known as general properties, are properties that both the carrier gas and analyte possess but to different degrees. Specific properties, such as detectors that measure nitrogen-phosphorous content, have limited applications but compensate for this by their increased sensitivity.

Each detector has two main parts that when used together they serve as transducers to convert the detected property changes into an electrical signal that is recorded as a chromatogram. The first part of the detector is the sensor which is placed as close the the column exit as possible in order to optimize detection. The second is the electronic equipment used to digitize the analog signal so that a computer may analyze the acquired chromatogram. The sooner the analog signal is converted into a digital signal, the greater the signal-to-noise ratio becomes, as analog signal are easily susceptible to many types of interferences.

An ideal GC detector is distinguished by several characteristics. The first requirement is adequate sensitivity to provide a high resolution signal for all components in the mixture. This is clearly an idealized statement as such a sample would approach zero volume and the detector would need infinite sensitivity to detect it. In modern instruments, the sensitivities of the detectors are in the range of 10^{-8} to 10^{-15} g of solute per second. Furthermore, the quantity of sample must be reproducible and many columns will distort peaks if enough sample is not injected. An ideal column will also be chemically inert and should not alter the sample in any way. Optimized columns will be able to withstand temperatures in the range of -200 °C to at least 400 °C. In addition, such a column would have a short linear response time that is independent of flow rate and extends for several orders of magnitude. Moreover, the detector should be reliable, predictable and easy to operate.

Understandably, it is not possible for a detector meet all of these requirements. The next subsections will discuss some of the more common types of gas chromatography detectors and the relative advantages and/or disadvantages of each.

Type of Detector	Applicable Samples	Detection Limit
Mass Spectrometer (MS)	Tunable for any sample	.25 to 100 pg
Flame Ionization (FID)	Hydrocarbons	1 pg/s
Thermal Conductivity (TCD)	Universal	500 pg/ml
Electron-Capture (ECD)	Halogenated hydrocarbons	5 fg/s
Atomic Emission (AED)	Element-selective	1 pg
Chemiluminescence (CS)	Oxidizing reagent	Dark current of PMT
Photoionization (PID)	Vapor and gaseous Compounds	.002 to .02 μg/L

Table: Typical gas chromatography detectors and their detection limits.

Mass Spectrometry Detectors

Mass Spectrometer (MS) detectors are most powerful of all gas chromatography detectors. In a GC/MS system, the mass spectrometer scans the masses continuously throughout the separation. When the sample exits the chromatography column, it is passed through a transfer line into the inlet of the mass spectrometer. The sample is then ionized and fragmented, typically by an electron-impact ion source. During this process, the sample is bombarded by energetic electrons which ionize the molecule by causing them to lose an electron due to electrostatic repulsion. Further bombardment causes the ions to fragment. The ions are then passed into a mass analyzer where the ions are sorted according to their m/z value, or mass-to-charge ratio. Most ions are only singly charged.

The Chromatogram will point out the retention times and the mass spectrometer will use the peaks to determine what kinds of molecules are exist in the mixture. The figure below represents a typical mass spectrum of water with the absorption peaks at the appropriate m/z ratios.

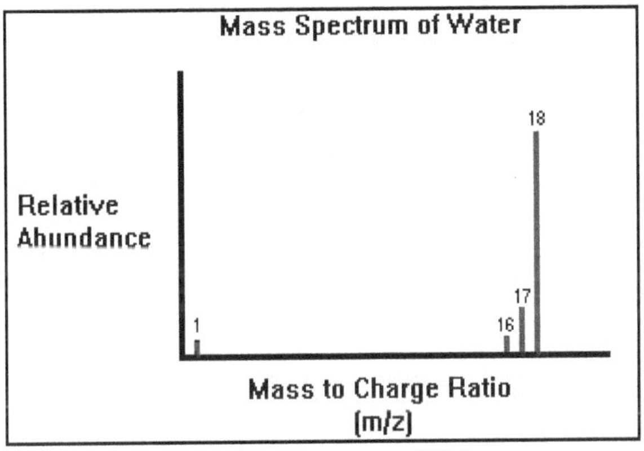

Figure: Mass Spectrum of Water.

Instrumentation

One of the most common types of mass analyzer in GC/MS is the quadrupole ion-trap analyzer, which allows gaseous anions or cations to be held for long periods of time by electric and magnetic fields. A simple quadrupole ion-trap consists of a hollow ring electrode with two grounded end-cap electrodes as seen in figure. Ions are allowed into the cavity through a grid in the upper end cap. A variable radio-frequency is applied to the ring electrode and ions with an appropriate m/z value orbit around the cavity. As the radio-frequency is increased linearly, ions of a stable m/z value are ejected by mass-selective ejection in order of mass. Ions that are too heavy or too light are destabilized and their charge is neutralized upon collision with the ring electrode wall. Emitted ions then strike an electron multiplier which converts the detected ions into an electrical signal. This electrical signal is then picked up by the computer through various programs. As an end result, a chromatogram is produced representing the m/z ratio versus the abundance of the sample.

GC/MS units are advantageous because they allow for the immediate determination of the mass of the analyte and can be used to identify the components of incomplete separations. They are rugged, easy to use and can analyze the sample almost as quickly as it is eluted. The disadvantages of mass spectrometry detectors are the tendency for samples to thermally degrade before detection and the end result of obliterating all the sample by fragmentation.

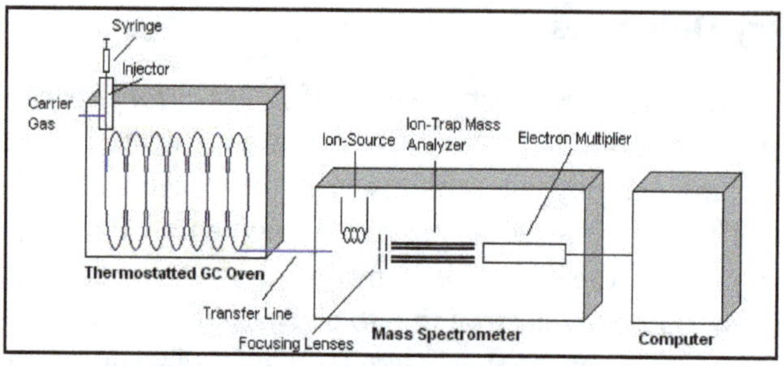

Figure: Schematic of the GC/MS system.

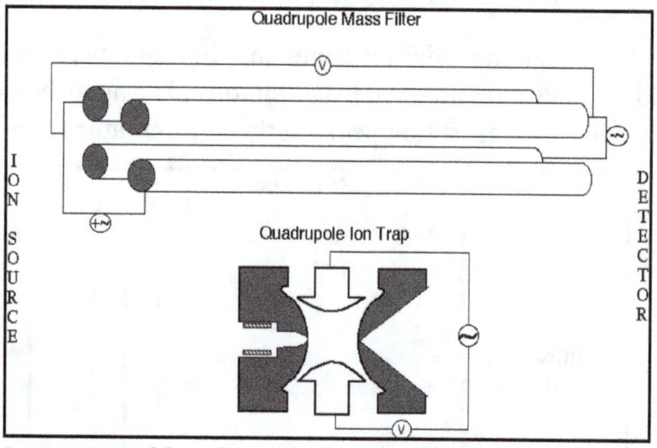

Figure: Arrangement of the poles in Quadrupole and Ion Trap Mass spectrometers.

Flame Ionization Detectors

Flame ionization detectors (FID) are the most generally applicable and most widely used detectors. In a FID, the sample is directed at an air-hydrogen flame after exiting the column. At the high temperature of the air-hydrogen flame, the sample undergoes pyrolysis, or chemical decomposition through intense heating. Pyrolized hydrocarbons release ions and electrons that carry current. A high-impedance picoammeter measures this current to monitor the sample's elution.

It is advantageous to used FID because the detector is unaffected by flow rate, noncombustible gases and water. These properties allow FID high sensitivity and low noise. The unit is both reliable and relatively easy to use. However, this technique does require flammable gas and also destroys the sample.

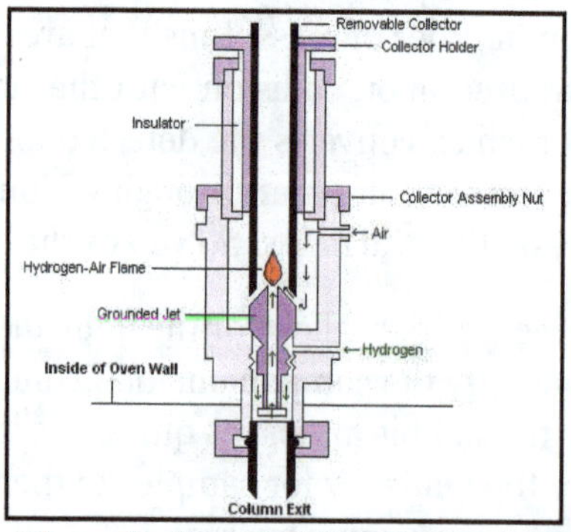

Figure: Schematic of a typical flame ionization detector.

Thermal Conductivity Detectors

Thermal conductivity detectors (TCD) were one the earliest detectors developed for use with gas chromatography. The TCD works by measuring the change in carrier gas thermal conductivity caused by the presence of the sample, which has a different thermal conductivity from that of the carrier gas. Their design is relatively simple, and consists of an electrically heated source that is maintained at constant power. The temperature of the source depends upon the thermal conductivities of the surrounding gases. The source is usually a thin wire made of platinum, gold. The resistance within the wire depends upon temperature, which is dependent upon the thermal conductivity of the gas.

TCDs usually employ two detectors, one of which is used as the reference for the carrier gas and the other which monitors the thermal conductivity of the carrier gas and sample mixture. Carrier gases such as helium and hydrogen has very high thermal conductivities so the addition of even a small amount of sample is readily detected.

The advantages of TCDs are the ease and simplicity of use, the devices' broad application to inorganic and organic compounds, and the ability of the analyte to be collected after separation and detection. The greatest drawback of the TCD is the low sensitivity of the instrument in relation to other detection methods, in addition to flow rate and concentration dependency.

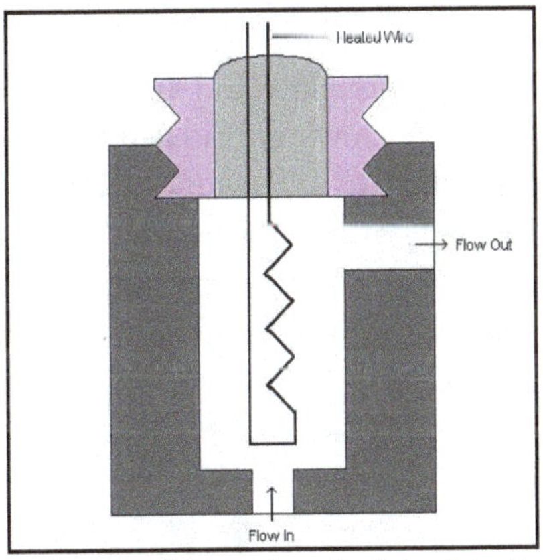

Figure:Schematic of thermal conductivity detection cell.

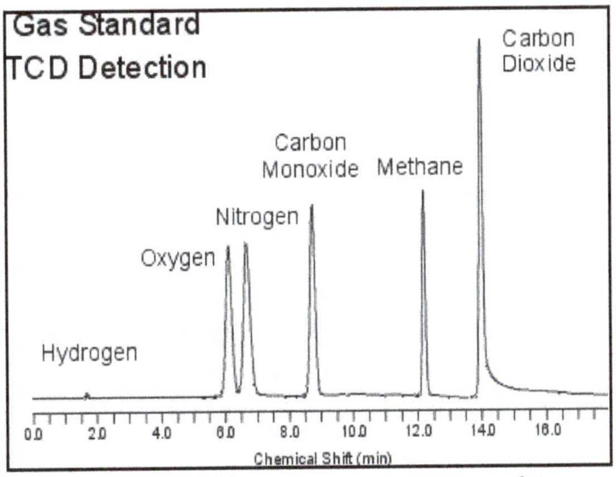

Figure: Standard Chromatogram of a Mixture of Gases.

Chromatogram

Figure represents a standard chromatogram produced by a TCD detector. In a standard chromatogram regardless of the type detector, the x-axis is the time and the y-axis is the abundance or the absorbance. From these chromatograms, retention times and the peak heights are determined and used to further investigate the chemical properties or the abundance of the samples.

Electron-capture Detectors

Electron-capture detectors (ECD) are highly selective detectors commonly used for detecting environmental samples as the device selectively detects organic compounds with moieties such as halogens, peroxides, quinones and nitro groups and gives little to no response for all other compounds. Therefore, this method is best suited in applications where traces quantities of chemicals such as pesticides are to be detected and other chromatographic methods are unfeasible.

The simplest form of ECD involves gaseous electrons from a radioactive emitter in an electric field. As the analyte leaves the GC column, it is passed over this emitter, which typically consists of nickle-63 or tritium. The electrons from the emitter ionize the nitrogen carrier gas and cause it to release a burst of electrons. In the absence of organic compounds, a constant standing current is maintained between two electrodes. With the addition of organic compounds with electronegative functional groups, the current decreases significantly as the functional groups capture the electrons.

The advantages of ECDs are the high selectivity and sensitivity towards certain organic species with electronegative functional groups. However, the detector has a limited signal range and is potentially dangerous owing to its radioactivity. In addition, the signal-to-noise ratio is limited by radioactive decay and the presence of O_2 within the detector.

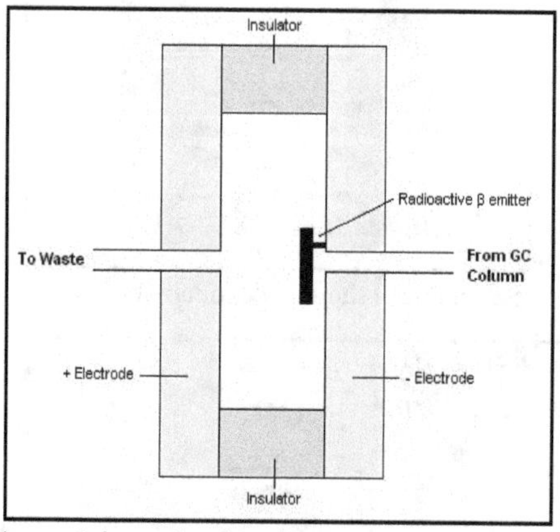

Figure: Schematic of an electron-capture detector.

Atomic Emission Detectors

Atomic emission detectors (AED), one of the newest addition to the gas chromatographer's arsenal, are element-selective detectors that utilize plasma, which is a partially ionized gas, to atomize all of the elements of a sample and excite their characteristic atomic emission spectra. AED is an extremely powerful alternative that has a wider applicability due to its based on the detection of atomic emissions. There are three ways of generating plasma: microwave-induced plasma (MIP), inductively coupled plasma (ICP) or direct current plasma (DCP). MIP is the most commonly employed form and is used with a positionable diode array to simultaneously monitor the atomic emission spectra of several elements.

Instrumentation

The components of the Atomic emission detectors include 1) an interface for the incoming capillary GC column to induce plasma chamber,2) a microwave chamber, 3) a cooling system, 4) a diffration grating that associated optics, and 5) a position adjustable photodiode array interfaced to a computer.

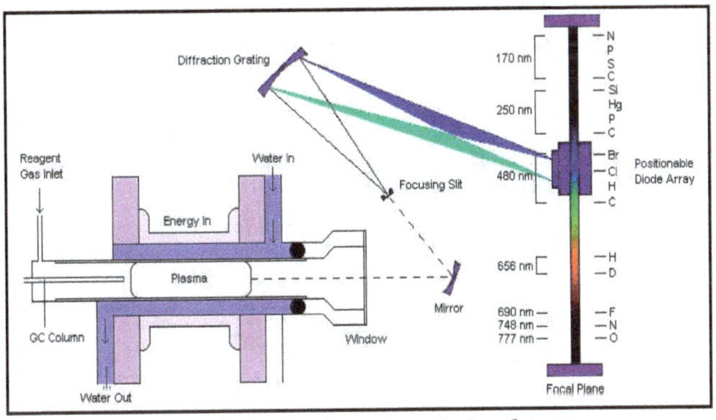

Figure: Schematic of atomic emission detector.

GC Chemiluminescence Detectors

Chemiluminescence spectroscopy (CS) is a process in which both qualitative and quantitative properties can be determined using the optical emission from excited chemical species. It is very similar to AES, but the difference is that it utilizes the light emitted from the energized molecules rather than just excited molecules. Moreover, chemiluminescence can occur in either the solution or gas phase whereas AES is designed for gaseous phases. The light source for chemiluminescence comes from the reactions of the chemicals such that it produces light energy as a product. This light band is used instead of a separate source of light such as a light beam.

Like other methods, CS also has its limitations and the major limitation to the detection limits of CS concerns with the use of a photomultiplier tube (PMT). A PMT requires a dark current in it to detect the light emitted from the analyte.

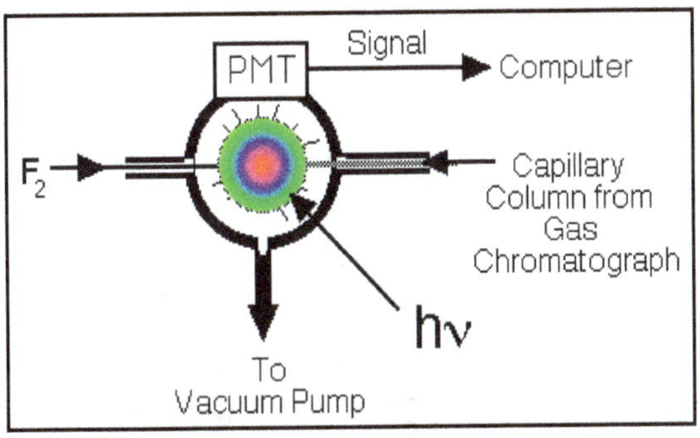

Figure: Schematic of a GC Chemiluminescence Detector.

Photoionization Detectors

Another different kind of detector for GC is the photoionization detector which utilizes the properties of chemiluminescence spectroscopy. Photoionization detector (PID) is a portable vapor and gas detector that has selective determination of aromatic hydrocarbons, organo-heteroatom, inorganice species and other organic compounds. PID comprise of an ultrviolet lamp to emit photons that are absorbed by the compounds in an ionization chamber exiting from a GC column. Small fraction of the analyte molecules are actually ionized, nondestructive, allowing confirmation analytical results through other detectors. In addition, PIDs are available in portable hand-held models and in a number of lamp configurations. Results are almost immediate. PID is used commonly to detect VOCs in soil, sediment, air and water, which is often used to detect contaminants in ambient air and soil. The disavantage of PID is unable to detect certain hydrocarbon that has low molecular weight, such as methane and ethane.

Instrumentation

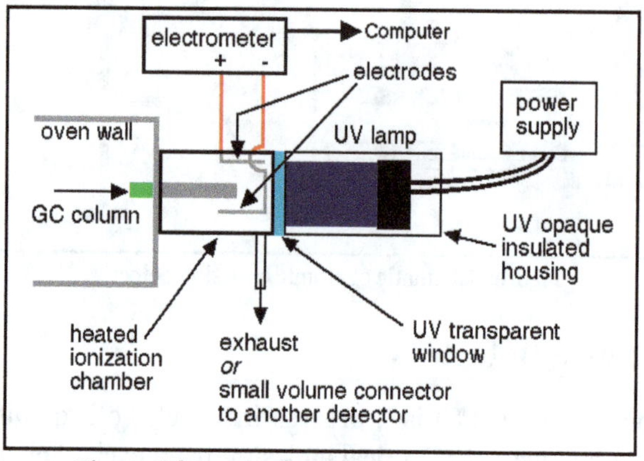

Figure: Schematic of a photoionization detector.

Limitations:

1. Not suitable for detecting semi-volatile compounds.
2. Only indicates if volatile organic compounds are presents.
3. High concentration so methane is required for higher performance.
4. Frequent calibration is required.
5. Units of parts per million range.
6. Environmental distraction, especially water vapor.
7. Strong electrical field's Rapid variation in temperature at the detector and naturally occurring compounds may affect instrumental signal.

Applications

Gas chromatography is a physical separation method in where volatile mixtures are separated. It can be used in many different fields such as pharmaceuticals, cosmetics and even environmental toxins. Since the samples have to be volatile, human breathe, blood, saliva and other secretions

containing large amounts of organic volatiles can be easily analyzed using GC. Knowing the amount of which compound is in a given sample gives a huge advantage in studying the effects of human health and of the environment as well.

Air samples can be analyzed using GC. Most of the time, air quality control units use GC coupled with FID in order to determine the components of a given air sample. Although other detectors are useful as well, FID is the most appropriate because of its sensitivity and resolution and also because it can detect very small molecules as well.

GC/MS is also another useful method which can determine the components of a given mixture using the retention times and the abundance of the samples. This method be applied to many pharmaceutical applications such as identifying the amount of chemicals in drugs. Moreover, cosmetic manufacturers also use this method to effectively measure how much of each chemical is used for their products.

Liquid Chromatography

Liquid chromatography (LC) is an analytical chromatographic technique that is useful for separating ions or molecules that are dissolved in a solvent.

If the sample solution is in contact with a second solid or liquid phase, the different solutes will interact with the other phase to differing degrees due to differences in adsorption, ion-exchange, partitioning, or size.

These differences allow the mixture components to be separated from each other by using these differences to determine the transit time of the solutes through a column.

Instrumentation Simple liquid chromatography consists of a column with a fritted bottom that holds a stationary phase in equilibrium with a solvent. Typical stationary phases (and their interactions with the solutes) are: solids (adsorption), ionic groups on a resin (ion-exchange), liquids on an inert solid support (partitioning), and porous inert particles (size-exclusion).

The mixture to be separated is loaded onto the top of the column followed by more solvent. The different components in the sample mixture pass through the column at different rates due to differences in their partitioning behavior between the mobile liquid phase and the stationary phase. The compounds are separated by collecting aliquots of the column effluent as a function of time.

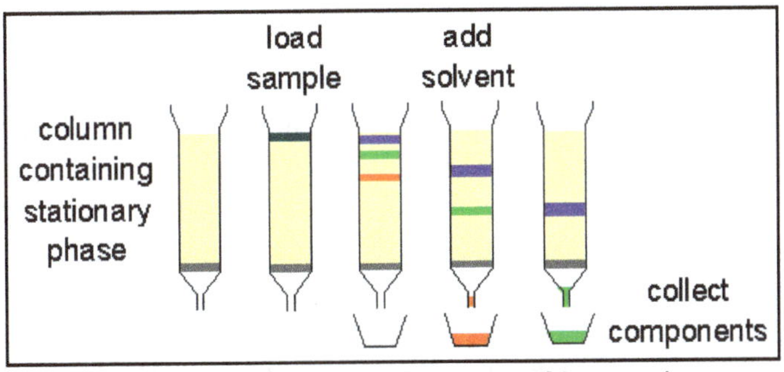

Schematic of a simple liquid chromatographic separation.

Conventional LC is most commonly used in preparative scale work to purify and isolate some components of a mixture. It is also used in ultra-trace separations where small disposable columns are used once and then discarded. Analytical separations of solutions for detection or quantification typically use more sophisticated high-performance liquid chromatography instruments.

Extraction

Solvent extraction technique is a part of analytical chemistry and has been recognized as an excellent separation method because of its ease, simplicity, speed, and wide scope. Utilizing apparatus no more complicated than a separately funnel, requiring just several minutes, at the most to perform, applicable both to trace and macro levels of metals, extraction procedures offers much to the analytical chemist. A further advantage of the extraction method over the widely used precipitation method lies in the cleaner separations that can be achieved by the former. With the later method contamination of precipitates by co-precipitation phenomena is a decided limitation which is minimized only with difficulty, whereas the analog of co-precipitation, i.e., co-extraction, is almost unknown in solvent extraction.

After world war second, chemists were engaged in atomic energy research, encountered the problem of separating and purifying almost all the elements, in amounts ranging from very low traces up to the usual micro levels. Among the solution to these difficult problems were the precipitation method and the ion exchange method. The former was found to be suitable for the treatment of materials in solution at high electrolyte concentrations and the later for the treatment of the ions in dilute solution or for the separation of chemically similar elements such as lanthanides and transplutonium elements.

However, the solvent extraction method proved to be most effective and attractive to separate these metals, because in this method, the separation is almost complete than precipitation method and operation time is very less as compared to ion exchange method. With proper choice of extracting agents, this technique can achieve group separation or selective separation of trace elements with high efficiencies. In analytical applications, solvent extraction may serve the following three purposes:

i) Pre-concentration of trace elements.

ii) Elimination of matrix interference.

iii) Differentiation of chemical species.

Solvent extraction or liquid-liquid extraction by high molecular weight organic amines has become increasingly popular in recent years in studying metal complexes. Extraction by these organic amines combine many of the advantages of both solvent extraction and ion exchange. The main interest of metal extraction by the high molecular weight amines lies in their selectivity towards anionic metal complexes, reversibly formed in an aqueous solution. Hence they are generally referred to as "liquid anion exchangers". The extent of extraction by the organic bases depends on their nature, structure, size, concentration and the nature of the organic solvent used as diluent.

1) The solubility of primary long chain amines in non-polar solvents increases with increasing chain length, secondary amines are generally highly soluble in non-polar solvents and

sparingly soluble in highly polar solvents. Tertiary amines are completely miscible with non-polar solvents at room temperature and sparingly soluble in polar solvents.

2) Extractive power of the alkylamines generally increases from primary to secondary to tertiary to quaternary amines. The trend may be exceptionally changed if there is extraction by bulky amines, whcrc stcric factor play an important role.

3) Third phase formation:

   a) The third phase formation, a second organic one, is more common in systems where the diluent is an aliphatic hydrocarbon. Aromatic and some derivatives of aliphatic hydrocarbons show the phenomenon usually at high organic phase loading.

   b) The splitting of organic phase is less common, when straight chain alkylamines are used.

   c) Formation of third phase is most common, characteristic of amine sulphate systems. The compatibility increases in the order: sulphate >bisulphate > chloride > nitrate.

   d) The formation of the third phase is temperature dependent. Literature in solvent extraction is reviewed by Freiser every two years in 'Analytical Chemistry'. Morrison and Freiser wrote a comprehensive monograph on solvent extraction in analytical chemistry. The solvent extraction of metal chelate complexes was reviewed by Starry and Zoltov. Marcus and Kertes and De et al, the applications were compiled by Sakine and Hasegawa. The extraction of many metal ions from various aqueous solutions by high molecular weight amines have been reviewed by Khopkar and Green.

High molecular weight amine systems used for liquid-liquid extraction of many metals from different aqueous solutions are listed as:

Adogen 364, Alaminc 336 (tricarpyl amine), Aliquat 336 (tricarpyl methyl ammonium chloride), Amerlite LA-1 (N-dodecyl trialkyl methyl amine), LA-2 (N-lauryl trialkyl methyl amine), n-octyl aniline, TBA (tribenzyl amine), TOA (tri-n-octyl amine), TIOA (tri-iso-octyl amine), TLA (tri-n-lauryl amine).

Basic Principles of Solvent Extraction Method

An extractant, is a substance primarily responsible for the transfer of a solute (here metal) from one phase to the other. The extractant is dissolved in a suitable diluent and together act as a solvent. The diluent is immiscible with other phase which is usually water. The extractant reacts with the solute by solvation/chelation/ion pair formation etc. to extract from the aqueous phase. The distribution equilibrium between two phases is governed by Gibbs phase rule, given by:

$$P+V=C+2$$

Where,

P = is the number of phases,

V = is the variance or degree of freedom,

C = is the number of components.

In solvent extraction, we have P=2 two phases namely aqueous and organic phase, the component C=1, viz. solute, in solvent and water phase and at constant temperature and pressure P=1, thus, we therefore have:

$$2+1=1+2 \qquad i.e. \qquad P+V=C+2$$

According to Nernst distribution law,

If $[X]_1$ is concentration of solute in phase 1 and $[X]_2$ is the concentration of solute in phase 2 at equilibrium:

$$K_D = \frac{[X]1}{[X]2}$$

Where K_D is called as the partition coefficient, this partition or distribution coefficient is independent of the total solute concentration in either of the phases. In the above expression for K_D, we have not considered the activity coefficient of the species in the organic as well as in the aqueous phase. We, therefore, use the term distribution ratio (D) to account for the total concentration of species in the two phases.

Distribution Ratio (D)

The distribution of a solute between two immiscible solvents in contact to each other can be described by the distribution ratio "D".

$$D = \frac{[X]_1}{[X]_2}$$

Where $[X]$ represents the stochiometric or formal concentration of a substance X and the subscripts 1 and 2 refer to the two phases. Since in most cases, two-phase system is of analytical interest, an organic solvent and aqueous are involved, D will be understood to be;

$$D = \frac{[X]_{org}}{[X]_{aq}}$$

The subscript org and aq refer to the organic and aqueous phases respectively. Distribution ratio 'D' is dimensionless quantity, separation of two solutes by solvent extraction is expressed by the term, separation factor (α), which is related to individual distribution ratios,

$$\alpha = \frac{D_A}{D_B}$$

D_A and D_B are the respective distribution ratios of solute A and B.

In the simplest extraction case, the distribution ratio is constant in accordance with the classical Nernst distribution law, a solute will distribute itself between two essentially immiscible solvents so that at equilibrium, the ratio of the concentrations of the solute in the two phases at a particular temperature will be constant, provided the solute is not involved in chemical interactions in either phases. For such a solute, then $K_D = D$. Now in practical work, instead of using term K_D or D one prefers to use term percentage extraction (E).

Percent Extraction (%E)

The more commonly used term for expressing the extraction efficiency by analytical chemist is the percent extraction "E", which is related to "D" as:

$$\%\text{Extraction}(E) = \frac{100D}{D + V_{aq} / V_{org}}$$

Where, V represent solvent volume and the other quantities remain as previously defined. The percent extraction may be seen to vary with the volume ratio of the two phases as well as with D.

Classification of Extraction Systems

The process of metal extraction is based on the formation of neutral metal chelate. All types of chelating agents find useful applications in metal extraction procedures. Various extraction systems can be classified in several ways. The classical one is based on the nature of the extracted species. The present day classification is based upon the process of extraction. Thus, based upon the process of extraction, extraction systems can be classified into four major classes viz.,

a) Chelate extraction.

b) Extraction by solvation.

c) Extraction involving ion pair formation.

d) Synergic extraction.

All thc above extractions are based on the fact that neutral or uncharged species are extracted easily in organic solvents. These are described here briefly:

Chelate Extraction

In this class, extraction proceeds by the process of formation of chelate or closed ring structure between the chelating agent and the metal ion to be extracted.

e.g. i) The extraction of Uranium with 8-hydroxyquinoline in chloroform.

ii) The extraction of Iron with cupferron in carbon tetrachloride.

Extraction by Solvation

In this class, the extraction proceeds by the process of solvation of the species which is extracted into organic phase. Oxygenated organic solvents such as alcohols (C-OH), ketones, ethers and esters show some basicity because of the lone pair of electron on the oxygen atom and can therefore directly solvate protons and metal ions and bring about their extraction.

E.g. i) The extraction of Uranium with tributyl phosphate from nitric acid.

ii) The extraction of Iron(III) with diethyl ether from hydrochloric acid.

Extraction involving ion pair formation

The extraction proceeds with the formation of neutral uncharged species which in turn gets extracted in to the organic phase. The best example of this is the extraction of Scandium and Uranium with trioctyl amine from mineral acids. In this case an ion pair is formed between complex of metal ion with high molecular weight amine and anionic species of mineral acids.

Synergic Extraction

In this case, there is enhancement in the extraction on account of use of two extractants.

e.g. the extraction of Uranium with tributylphosphate (TBP) as well as 2- thionyltrifluroacetone (TTA).

Methods of Extraction

Three basic methods of liquid-liquid extraction are generally utilized in the analytical laboratory.

1. Batch Extraction

Batch extraction, the simplest and most commonly used method, consists of extracting the solute from one immiscible layer in to other by shaking the two layers until equilibrium is attained, after which the layers are allowed to settle before sampling. This is commonly used on the small scale in chemical laboratories. The most commonly employed apparatus for performing a batch extraction is a separatory funnel. The batch extractions may also be used with advantage when the distribution ratio is large.

2. Continuous Extraction

The second type, continuous extraction, makes use of a continuous flow of immiscible solvent through the solution or a continuous countercurrent flow of both phases. Continuous extractions are particularly applicable when the distribution ratio is relatively small. Continuous extraction device operate on the same general principle, which consist of distilling the extracting solvent from a boiler flask and condencing it and passing it continuously through the solution being extracted. The extracting liquid separates out and flows back into the receiving flask, where it is again evaporated and recycled while the extracted solute remains in the receiving flask. When the solvent cannot easily be distilled, a continuous supply of fresh solvent may be added from a reservoir.

3. Countercurrent Extractions

Extraction by continuous countercurrent distribution is the third general type and is used primarily for fractionation purposes. The separation through continuous countercurrent method is achieved by virtue of the density difference between the fluids in contact. In vertical columns, the denser phase enters at the top and flows downwards while the less dense phase enters from the bottom and flows upwards. The choice of method to be employed will depend primarily upon the value of the distribution ratio of the solute of interest, as well as on the separation factors of the interfering materials.

Factors Influencing the Extraction Efficiency

Primary requirement of solvent extraction for separation /removal purposes is a high distribution ratio of the solute of interest between the two liquid phases. It is useful to employ a number of different techniques for enhancing the distribution ratio. It depends on the nature of the species being extracted and extraction system. The attainment of selectivity in an extraction procedure is also very important. Some of the factors, which affect the distribution of solute of interest, are given below.

A. Choice of Solvent

The most important consideration in the selection of a solvent for use in a particular extraction procedure is the extractability of the element of interest. For subsequent analytical processing a consideration of the solubility of the solute in particular solvent, the ease of recovery of the solvent or the ease of recovery of the solute from the solvent is very important. Thus, the boiling point of the solvent or the ease of stripping by chemical reagents enters into selection of a solvent when the possibility of a choice existed. Similarly, the degree of miscibility of the two phases, the relative specific gravities, viscosities and tendency to form emulsion should be considered. From the point of view of safety, the toxicity and the flammability of the solvent must be considered.

Use of a suitable solvent for effective separation is very important. Metal chelates and many organic molecules, being essentially covalent compounds do not impose many restrictions on the solvent and the general rules of solubility are the great use. In ion association systems and particularly in oxonium type ions, the role of solvents is very important. This is due to involvement of solvent in the formation of extractable species.

B. Acidity of an Aqueous Phase

The extractability of metal complexes is greatly influenced by the acidity of an aqueous phase, so it is necessary to assure optimum concentration of H+ ions for maximum extraction. In the case of chelate extraction, the chelating reagent concentration is maintained constant; the distribution of the metal in a system is a function of pH. For this reason, curves of extractability versus pH at constant reagent concentration are of great analytical significance. Sometimes it is possible to achieve the desired characteristics of a solvent by employing a mixed solvent system.

C. Stripping

Stripping is the removal of the extracted solute from the organic phase for further processing or analysis. In many colorimetric procedures and even radioactive techniques, the concentration of solute is determined directly in the organic phase. However, where further separation steps are required, it is necessary to remove the solute from the organic layer to more stable medium. When organic layer is on the steam bath, care should be taken, to avoid loss of volatile solute during evaporation. Addition of acid to water before evaporation helps to break the chelate complexes, thereby causing the metal ion to enter the aqueous phase. In the process of destroying the residual organic matter, hydrochloric acid, nitric acid, perchloric acid or aqua regia is used.

The usual procedure is to shake the organic layer with a volume of water alone or water containing an appropriate concentration of acid, an oxidizing or reducing agent or masking agent. The metal ion is then back extracted in the stripping aqueous phase.

The conditions employed depends upon metal ion and the particular extraction system and are such that they promote the reversal of extraction. Pure water or water adjusted to an appropriate pH/ molarity of acid are the more popular and convenient stripping agents. Washing the organic layer with an oxidizing and reducing agent changes the metal ion to be stripped in an oxidation state in which it is not extracted under the specific conditions.

D. Use of Masking Agents

In the extraction procedures for metal pairs that are difficult to separate; masking or sequestering agents are introduced to improve the separation factor. Masking agents are themselves metal-complexing agents, which serve to prevent a particular metal from taking part in their usual reaction, and thus to remove their interference without the necessity of actual separation. In solvent extraction, masking agents are used to prevent certain metals from forming extractable complexes and increasing the selectivity of the method. In extraction of metals the application of masking agents, which include cyanide, tartarate, citrate, fluoride, and EDTA. The selection of a particular masking agent largely depends on the acidity of the system and the complex formation constants of the metal with both the masking and the extraction agent.

EDTA, which has been proved as most useful masking agent, forms anionic complexes with a number of metal ions.

E. Salting-out Agents

The term salting-out agent is applied to those electrolytes whose addition greatly enhances the extractability of complexes. The function of salting-out agent would be primarily of providing a higher concentration of complex and thus improve the extraction. Water is probably bound as a shell of oriented water dipoles around the ion and thus becoming unavailable as "free solvent".

Addition of salting-out agents decreases the dielectric constant of the aqueous phase, which favors the formation of the ion association complexes. Salting-out agents have been used with great success in separation involving the halide and thiocynate systems.

In addition to enhancement of the extraction of the metal of interest using saltingout agents, it is also possible to decrease the extraction of impurities in the system. Thus, it is necessary to choose an agent that produces a favorable separation factor between the element of interest and the impurities. However, it must be remembered that anomalies sometimes result from specific interaction effects. Aluminium or calcium salts are strong salting-out agents, whereas ammonium salts are much weaker but analytically more convenient.

F. Backwashing

Backwashing is an auxiliary technique used with batch extractions to influence quantitative separations of elements. The combined organic phases from several extractions of the original aqueous phase contain practically all the desired elements and possibly some of the impurities that have been extracted to a much smaller extent. This combined organic phase when shaken with one or more small portions of a fresh aqueous phase containing the optimum reagent/salting agent concentration, acidity, etc., will result in a redistribution of the impurities in favor of the aqueous phase since their distribution ratios are low. Under optimum conditions, most of the elements of interest will remain in the organic layer, since their distribution ratios are high.

This technique is analogous in many respects to the re-precipitation step in a gravimetric precipitation procedure. With the proper conditions, most of the impurities can be removed by this backwashing operation, with neglisible loss of the main component, thereby attaining a selective operation.

G. Variation of Oxidation State

The selectivity of an extraction is increased by the modification of oxidation states of the interfering ions present in solution, in order to prevent the formation of their extractable metal complexes e.g. reduction of Cerium(IV) to Cerium(III) prevents extraction of this element from nitrate media, the extraction of Iron(III) from chloride solutions can be prevented by reduction to Iron(II), which is not extractable. Similarly, Antimony(V) may be reduced to the tetravalent state to suppress its extraction. Conversely, it is important in the preparation of a solution for extraction to adjust the proper valence state of metal ion required for the formation of the complex in order to ensure complete extraction of that element. Selectivity can also be achieved by variation of the oxidation state of the co-extracted interfering ions during the stripping operation.

H. Synergic Extraction

Synergism is defined as the combined action of two complexing reagents, which is greater than the sum of the actions of the individual reagents used alone. An example of the synergic extraction of Ce(III) with picrolonic acid and benzo-15-crown-5.

I. Use of Organic Acid Media

Organic acid media are having ability of controlling the concentration of the complexing ligand, is one of the unique application, the ease of adjustment of pH and the wide difference in pH at which various metal ions form anionic complexes. The comparative ease of stripping of the complexes from the organic phase can be achieved by fully exploiting the differences in reactivity of various metals to backwash in the aqueous phase by mineral acid. It is known that organic acid media offers better separation of metals possibly due to high stability of metal organic acids complexes.

Solid-phase Extraction

SPE is a sample preparation technology that uses solid particle, chromatographic packing material, usually contained in a cartridge type device, to chemically separate the different components of a sample. Samples are nearly always in the liquid state [although specialty applications may be run with some samples in the gas phase]. Figure shows a sample, which appears black, being processed on a SPE device so that the individual dye compounds, which make up the sample, are chromatographically separated.

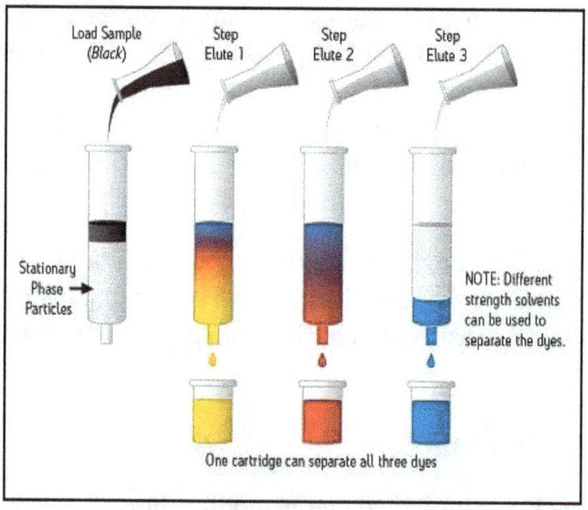

Figure: Examples of an SPE Method.

The chromatographic bed can be used to separate the different compounds in a sample, to make subsequent analytical testing more successful. For example, SPE is often used for the selective removal of interferences.

The technically correct name for this technology is "Liquid-Solid Phase Extraction," since the chromatographic particles are solid and the sample is in the liquid state. The same basic chromatographic principles of liquid chromatography that are used in HPLC are also used here, but in a different format and for a different reason. Here, chromatography is used to better prepare a sample before it is submitted for analytical testing.

In sample preparation, samples can come from a wide range of sources. They can be biological fluids such as plasma, saliva, or urine; environmental samples such as water, air, or soil; food products such as grains, meat, and seafood; pharmaceuticals; nutraceuticals; beverages; or industrial products. Even mosquito heads can be the sample. When a scientist needed to analyze neuropeptides extracted from the brains of mosquitoes, SPE was the sample preparation method of choice.

Four Major Benefits of SPE

There are many benefits to using SPE, but four major benefits deserve special attention.

1. Simplification of Complex Sample Matrix along with Compound Purification

One of the most difficult problems for an analytical chemist is when compounds of interest are contained in a complex sample matrix, such as mycotoxins in grains, antibiotic residues in shrimp, or drug metabolites in plasma, serum, or urine. The large number of interfering constituents or substances in the sample matrix along with the compounds of interest makes analysis extremely difficult.

The first problem to solve is the resulting complexity of the analysis itself due to the presence of so many entities which must be separated in order to identify and quantitate the compound[s] of interest.

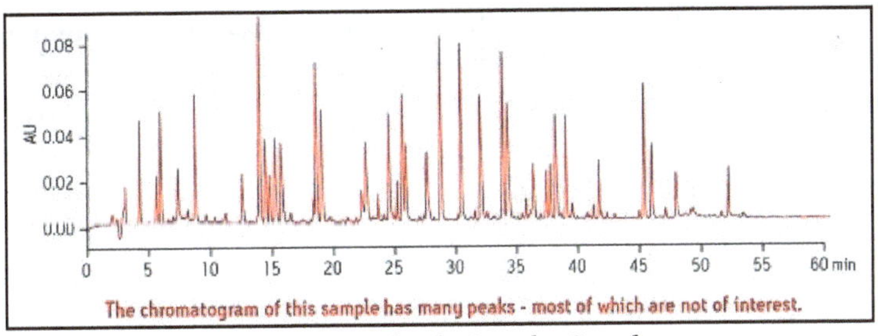

Figure: Example of a Complex Sample.

The robustness of the assay may not be adequate, because any slight change could impact the resolution of the separation of a critical pair of analytes.

Another consideration is that the presence of all the interferences in the original sample matrix can result in instrument downtime due to a buildup of contamination with each injection. If the interferences were removed as part of the sample preparation, then the compounds of interest could be analyzed with a simpler, more robust method. This can be seen in figure, which compares the original sample on top to the new SPE-prepared sample on the bottom.

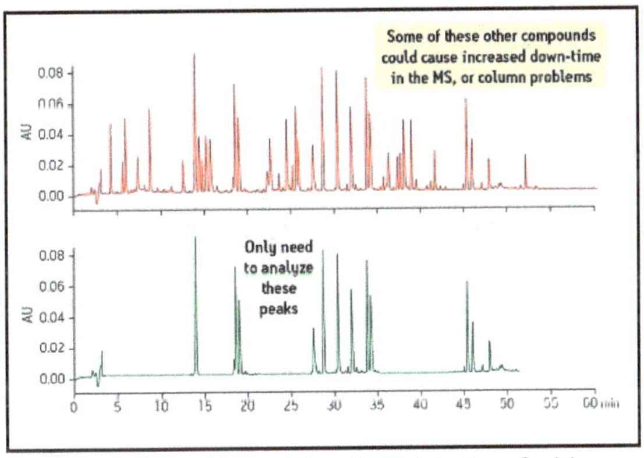

Figure: Comparison of Sample Matrix Complexities.

An additional benefit of simplifying the sample matrix is improved quantitation accuracy. The top blue trace for compound 1 in figure, initially appears to be acceptable. However, it really has some contamination from the sample matrix when compared to the blank sample matrix trace in red shown just beneath it. With a proper SPE protocol, the lower traces show the same compounds with no problems with interference, making quantitation much more accurate.

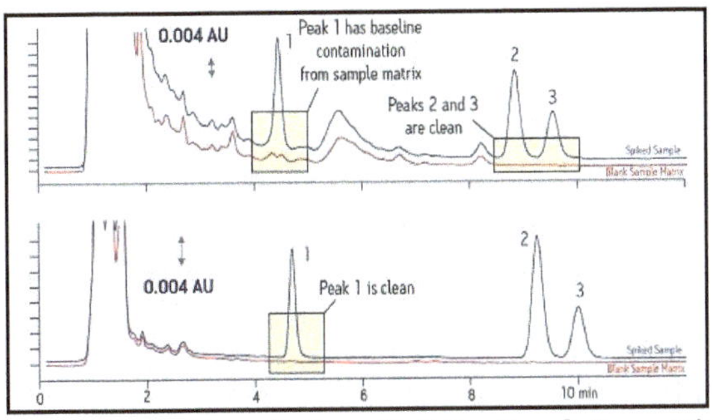

Figure: Improved Quantitation with Better Sample Preparation.

Another example is shown in figure. The upper trace shows significant interference from the sample matrix on both compounds 1 and 2. The lower trace shows much improved results [clean base line] due to proper sample preparation with SPE. Notice a much cleaner baseline improves the accuracy of the analytical results. Also, a much purer extract can be obtained if the sample requires isolation and purification of that compound.

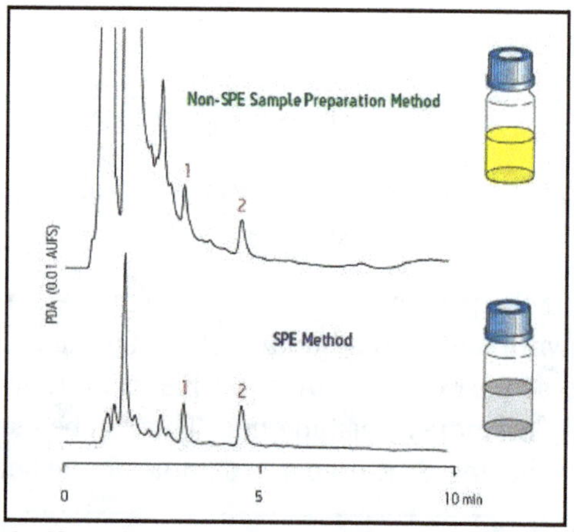

Figure: Significant Improvement in Baseline Using SPE Technology.

1. Reduce Ion Suppression or Enhancement in MS Applications

The second problem with complex sample matrices can be seen when we look at mass spectrometer output [LC/MS or LC/MS/MS]. For proper MS signal response [sensitivity], the compound ion must be allowed to form properly. In cases where the formation of the compound ion is suppressed by interferences in the sample matrix, the signal strength is greatly diminished.

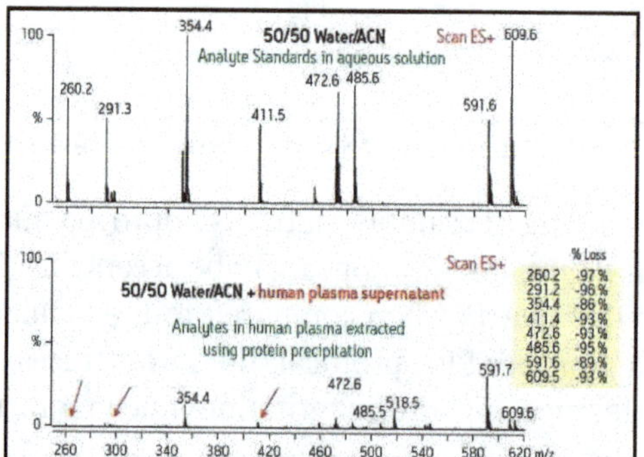

Figure: Example of Ion Suppression Due to Sample Matrix.

We can see this effect in figure. The upper output is the signal for our compounds of interest when injected in a saline solution. The lower trace shows significant reduction in response [> 90% suppression] of these same compounds when they were analyzed in human plasma. For the lower trace, only a common protein precipitation step was performed. This technique does not clean up the matrix interferences that cause the ion suppression, resulting in poor signal response.

Another good example of this suppression effect can be seen in figure. In the upper trace of the MS output, where the plasma sample was prepared with just a protein precipitation step, we can see that the terfenadine peak is suppressed by 80%. In the lower trace, where the same sample was prepared with a SPE method, we can see minimal ion suppression. Because the interferences from the sample matrix were removed, this allowed the compound ion to form properly, creating a better signal.

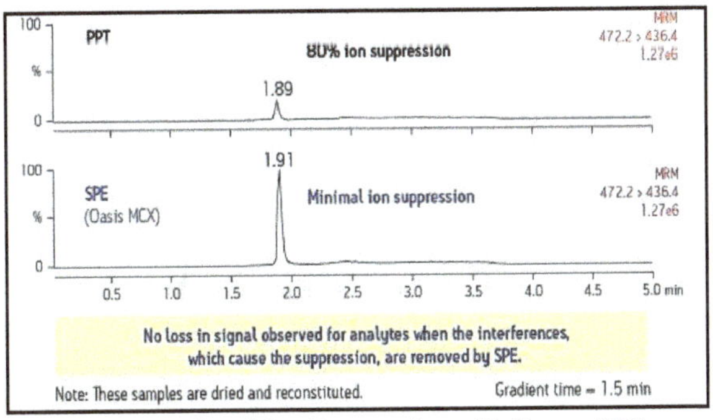

Figure: Reduced Ion Suppression with Proper SPE

In some instances, interferences from the sample matrix can artificially increase the signal reported for a compound. This is called ion enhancement, resulting in an inaccurately high reported value. A proper SPE method will minimize this effect by cleaning away the interferences from the compound, resulting in a more accurate reported value.

2. Capability to Fractionate Sample Matrix to Analyze Compounds by Class

An analyst may be faced with a sample that contains many compounds, with a need to separate them by class so that further analysis can be carried out much more efficiently. For example, a soft drink beverage contains a wide range of compounds in its formulation. An SPE method could be developed to separate the different classes of compounds, for example by their polarity. The polar compounds could be collected, as a separated fraction, from the more non-polar compounds. These two fractions could then be separately analyzed in a much more efficient way because their compounds would be more similar.

An example of the power of fractionation by SPE is shown in figure below. Here, a complex sample of a dry powder [purple grape drink mix] is easily separated into four fractions: a fraction of just the polar compounds, a purified red compound, a purified blue compound, and a fraction containing all the remaining very non-polar compounds.

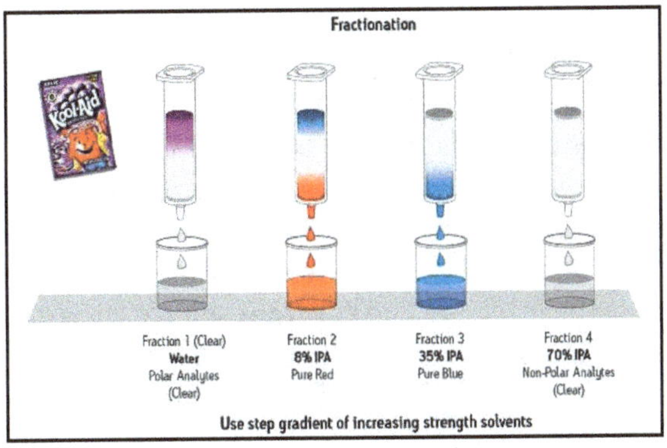

Figure: Sample Preparation by SPE.

Trace Concentration [Enrichment] of very Low Level Compounds

Analysts today often need to report on compounds at far lower concentration levels than ever before, as little as parts per trillion [ppt] and even lower. Typically these levels are lower in the neat sample than the sensitivity capability of the analytical instruments.

A good example of this is the analysis for trace contaminants in environmental samples or metabolite development over time in biological fluids. The upper trace in figure shows the poor response of the original neat sample for the compound of interest. Using the same analytical conditions but with the sample prepared with SPE used in a trace concentration strategy, the lower trace shows a dramatic increase in signal strength for this compound. With this result, an accurate calculation of the original compound concentration in the neat sample can be made.

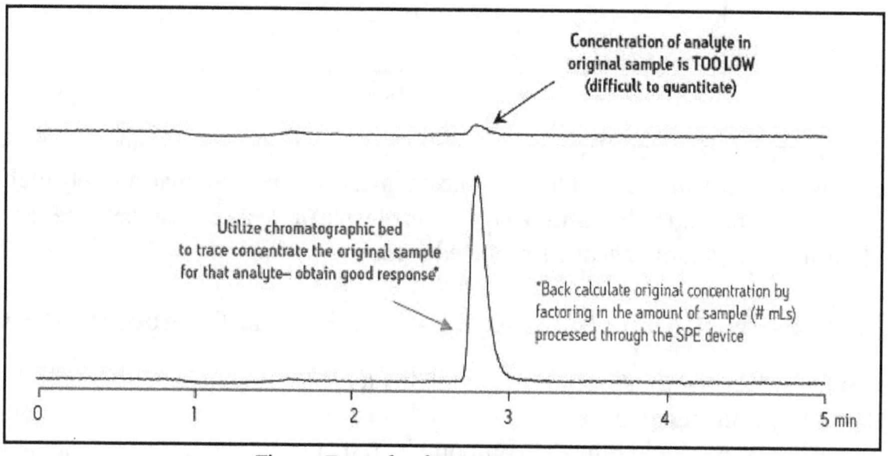

Figure: Example of Trace Concentration.

Without the retention capability of chromatographic packing materials in SPE, the ability to trace concentrate a specific compound[s] would be very difficult, if not impossible, with other sample preparation approaches.

Filtration

Filtration is one of the process of separation and it is defined as the action or process of filtering something. The act or process of filtering, especially the process of passing a liquid or gas, such as air, through a filter in order to remove solid particles. In liquid filtration Journals, such as that applied in waste material treatment, a liquid can be pulled through the filter by attraction, as in the examples given. On the opposite hand, some sort of applied pressure, or a pressure differential created by the existence of a vacuum, can force the liquid through the filter. Gas filtration is employed in a common appliance, the vacuum cleaner, which passes a stream of dust-filled air through a filtering bag inside the machine. The bag traps solid particles, while allowing clean air to pass back out into the room. This is essentially the same principle applied in air filters and even air conditioning and heating systems, which, in addition to regulating temperature, also remove dust, pollen, and other impurities from the air.

Filtration Techniques

Gravity Filtration

Gravity filtration uses a polyethylene or glass funnel with a stem and filter paper. Filter paper can have pore sizes ranging from small to large to permit slow to fast filtering. The paper is folded in half, then folded in quarters, and the tip of one corner is torn off to allow for a snug fit in the funnel cone. (If the paper has been pre-weighed, the torn corner piece must be saved to add to the post-filter weighing to avoid any errors.) The paper cone is fitted to the funnel so three thicknesses of the paper line one-half of the cone and one thickness lines the opposite half. Now place the funnel into a beaker and wet the filter paper completely with the dominate solvent or solvents in the mixture to be filtered.

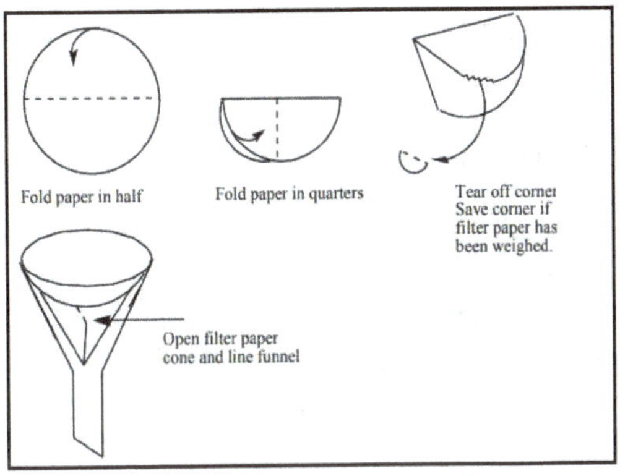

Figure: How to fold filter paper.

This step adheres the filter paper to the funnel walls preventing solid from escaping. Then, support the funnel with a clamp or ring (if necessary) and place a clean beaker beneath the funnel so the stem rests against the side of the beaker (this prevents splattering).

Before filtering, allow most of the solid in the mixture to settle. Now pour the supernatant liquid (the liquid standing over the solid in a mixture) through the filter first. This will allow the initial part of the filtration to proceed faster and may prevent clogging of the filter by the solid. To prevent splattering pour the liquid down a glass rod as shown in figure.

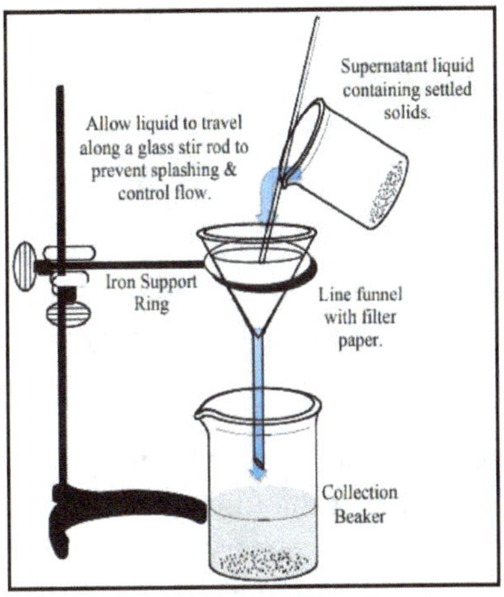

Figure: Gravity Filtration.

Scrape the solid onto the filter with a rubber policeman or spatula. Rinse the spatula, glass rod and beaker and pour the washings into the filter funnel. If the remaining solid residue is to be washed, rinse with three small portions (a few milliliters each) of an appropriate solvent. If the solid is to be saved, remove the filter paper carefully and place it on a watch glass to dry. Caution: Wet filter paper tears easily.

Sometimes the filtrate is tested to determine if a product or reactant has or has not passed through the filter. The test depends on what is being separated. For example: All barium ions $\left(Ba^{2+}\right)$ should have removed from solution by the formation of the precipitate $BaSO_4$. To check this a few drops of Na_2SO_4 solution can be added to a small portion of the filtrate. If no $BaSO_4$ precipitate forms, the filtration was successful. If a precipitate forms, additional precipitating (sulfate ion containing) reagents must be added to the filtrate and the resulting mixture must be filtered again.

Vacuum (or Suction) Filtration

Vacuum filtration uses a Buchner (pronounced "byook-ner") funnel and a water aspirator assembly. A Buchner funnel is a flat bottomed, porous, circular porcelain bowl with a short stem. The stem is fitted with a rubber stopper and inserted in the mouth of a side arm filter flask. Circular filter paper, the same diameter as the bowl, is placed on the flat bottom and wetted with the appropriate solvent to create a seal before starting the filtration.

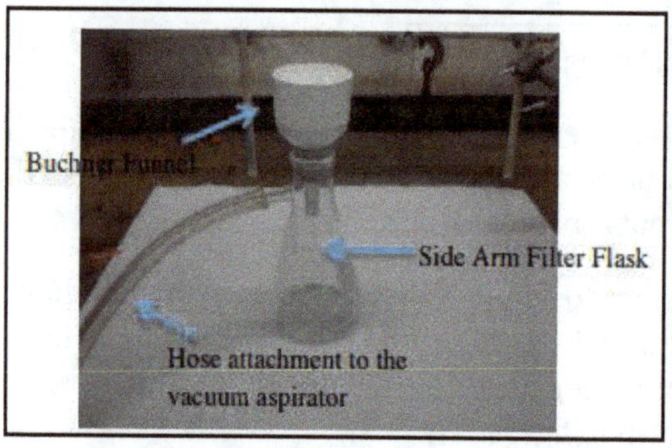

Figure: Buchner Funnel on a side arm filter flask.

Figure: Vacuum Aspirator (front & rear views).

The hose in the above figure is attaches the side arm of the filter flask to a vacuum aspirator in the hood or at the lab bench. This vacuum aspirator creates the suction that pulls liquid through the filter and filter paper.

Figure shows the entire set-up consisting of a Buchner funnel, the side arm flask, and the vacuum aspirator. When the water aspirator is turned on the flow of water creates suction. The hose is part of the filtration system and should be heavy enough to prevent pinching or collapse under external atmospheric pressure.

To filter a sample, turn on the aspirator and carry out the filtration in the same manner described for gravity filtration. (Note: NEVER pry off the funnel if the system is under vacuum! Water can flash back into the collection flask or the filter paper can be damaged resulting in the loss of filtered solid.) Turn off the water aspirator before carefully removing the wet filter paper without tearing.

Reaction byproducts (either the solid or filtrate) should be placed into appropriate labeled containers in the hood.

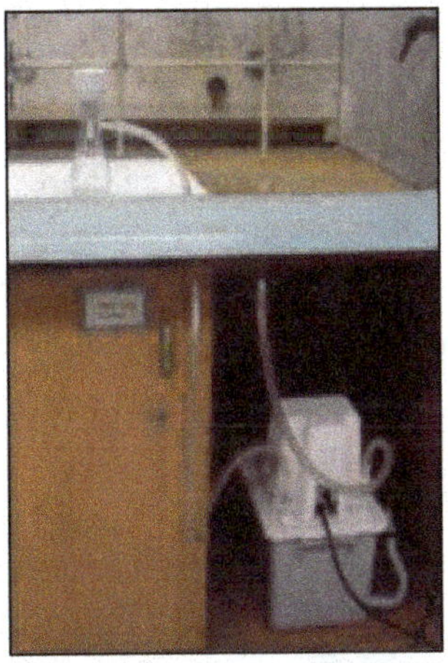

Figure: Vacuum Filtration Assembly.

Decantation

Decantation is a process to separate mixtures by removing a liquid layer that is free of a precipitate. The purpose may be to obtain a decant (liquid free from particulates) or to recover the precipitate. Decantation relies on gravity to pull precipitate out of solution, so there is always some loss of product, either from precipitate not fully falling out of solution or from leaving some liquid when separating it from the solid portion.

A piece of glassware called a decanter is used to perform decantation. There are several decanter designs. A simple version is a wine decanter, which has a wide body and a narrow neck. When wine is poured, solids stay in the base of the decanter. In the case of wine, the solid is usually potassium bitartrate crystals. For chemistry separations, a decanter may have a stopcock to drain away the precipitate or dense liquid or may have a partition to separate fractions.

Working of Decanting

There are two main decantation methods:

Separating Liquids and Solids

Decanting is done to separate particulates from a liquid by allowing the solids to settle to the bottom of the mixture and pouring off the particle-free part of the liquid.

For example, a mixture (possibly from a precipitation reaction) is allowed to stand so that gravity has time to pull the solid to the bottom of a container. The process is called sedimentation. Using gravity only works when the solid is less dense than the liquid. Clearwater may be obtained from mud by simply allowing time for the solids to separate from the water.

The separation may be enhanced using centrifugation. If a centrifuge is used, the solid may be compacted into a pellet, making it possible to pour off the decant with minimal loss of liquid or solid.

Separating Two or more Liquids

Another method is to allow two immiscible liquids to separate and the lighter liquid is poured or siphoned off. A common example is decantation of oil and vinegar. When a mixture of the two liquids is allowed to settle, the oil will float on top of the water so the two components may be separated. Kerosene and water can also be separating using decantation.

The two forms of decantation may be combined. This is especially useful if it's important to minimize loss of a solid precipitate. In this case, the original mixture may be allowed to settle or may be centrifuged to separate the decant and sediment. Rather than immediately drawing off the liquid, a second immiscible liquid may be added that is denser than the decant and that doesn't react with the sediment. When this mixture is allowed to settle, the decant will float on top of the other liquid and sediment. All of the decant can be removed with minimal loss of precipitate (except a tiny amount that remains floating in the mixture). In an ideal situation, the immiscible liquid that was added has a high enough vapor pressure that it evaporates, leaving all of the sediment.

Examples of Mixtures that can be Decanted

Some mixtures that can be decanted:

- Oil and water: Oil floats on top of water. Decanting the mixture allows the oil to be poured off the water.
- Gasoline or kerosene and water: This mixture is an example often cited as a safety hazard. Decanting a mixture containing flammable solvents can be dangerous as the flammable material evaporates and forms dangerous fumes.
- Dirt and water: Muddy water can be cleared up by decanting. The soil will sink to the bottom of the tube allowing the clear water to be poured off.
- Wine: Sediment from the fermentation process can produce an undesirable taste. Wine is decanted to separate the wine from these sediments.
- Cream and milk: Cream is separated from milk by decantation. Cream rises to the top of the milk mixture and is easily skimmed off.
- Blood and plasma: A centrifuge is necessary for this decantation. Plasma can be removed from blood by decantation.

Sublimation

Sublimation separates a mixture of solids, one of which sublimes. A few substances change directly, from a solid to vapor on heating without going through the liquid state. This change is called sublimation. On cooling, water vapor changes back to solid directly.

Iodine is a solid that sublimes. if a mixture of iodine and sand is heated in a beaker, the iodine changes from solid to vapor directly. This is a picture of sublimation process.

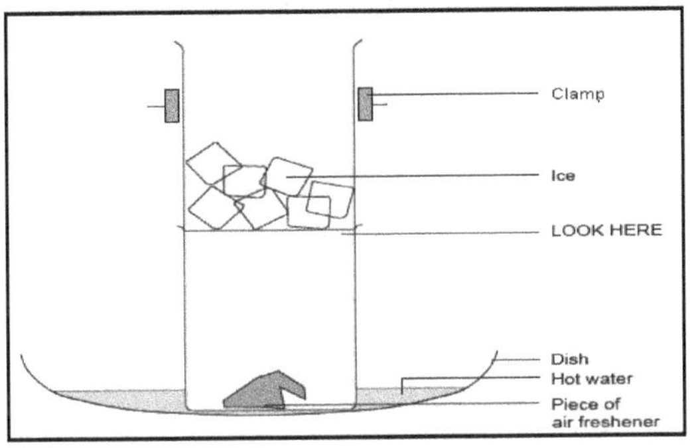

The vapor changes back to solid directly on a cold surface. The sand is not affected by the heat and remains in the beaker. Other substances that sublime include carbon dioxide, anhydrous iron (III) chloride, anhydrous aluminium chloride and some ammounium compounds. As only a few solids sublime, this method of separation is limited.

Examples of Sublimation

- Naphthalene: When used as insecticide and put between clothes, we can notice that it loses weight gradually. This loss of weight is due to sublimation. Naphthalene ball converts slowly to gas on standing. This gas is toxic to insects and hence they stay away from clothes.
- Water: Even water shows this behavior at triple point.
- Iodine: Iodine is a good example for sublimation.

Uses of Sublimation

- Drying: Sublimation helps to dry substance within getting moist. This process of sublimation is also beneficial in medicine. It is used in making powders from heat sensitive materials etc.
- Purification: If is a substance is in contaminated form, it can be separated into pure form by sublimation technique. Since the sublimation temperature of substance is different than the contaminant, when sublimation is done, the substance is separated leaving the contaminant behind.
- Aroma: Sublimation is the principle involved in air freshener solids used in room and toilets. The room fresheners are solids packed in a box. They spread the odor to the entire room by getting evaporated.

Evaporation

A puddle of water left undisturbed eventually disappears. The liquid molecules escape into the gas phase, becoming water vapor. Vaporization is the process in which a liquid is converted to a gas.

Evaporation is the conversion of a liquid to its vapor below the boiling temperature of the liquid. If the water is instead kept in a closed container, the water vapor molecules do not have a chance to escape into the surroundings and so the water level does not change. As some water molecules become vapor, an equal number of water vapor molecules condense back into the liquid state. Condensation is the change of state from a gas to a liquid.

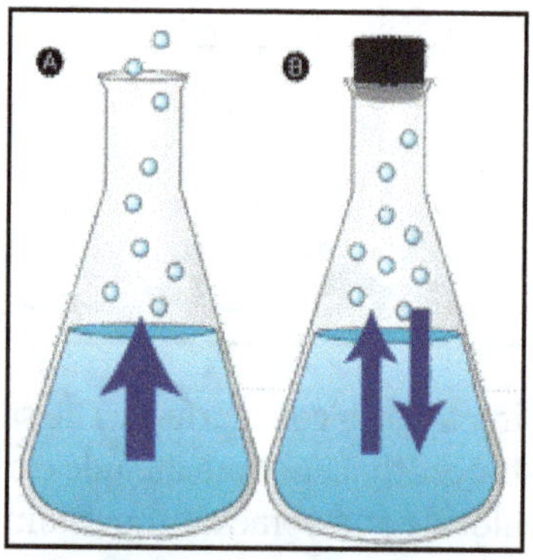

Figure: *Evaporation (A) and condensation (B).*

In order for a liquid molecule to escape into the gas state, the molecule must have enough kinetic energy to overcome the intermolecular attractive forces in the liquid. Recall that a given liquid sample will have molecules with a wide range of kinetic energies. Liquid molecules that have this certain threshold kinetic energy escape the surface and become vapor. As a result, the liquid molecules that remain now have lower kinetic energy. As evaporation occurs, the temperature of the remaining liquid decreases. You have observed the effects of evaporative cooling. On a hot day, the water molecules in your perspiration absorb body heat and evaporate from the surface of your skin. The evaporating process leaves the remaining perspiration cooler, which in turn absorbs more heat from your body.

A given liquid will evaporate more quickly when it is heated. This is because the heating process results in a greater fraction of the liquid's molecules having the necessary kinetic energy to escape the surface of the liquid. The figure below shows the kinetic energy distribution of liquid molecules at two temperatures. The numbers of molecules that have the required kinetic energy to evaporate are shown in the shaded area under the curve at the right. The higher temperature liquid (T_2) has more molecules that are capable of escaping into the vapor phase than the lower temperature liquid (T_1).

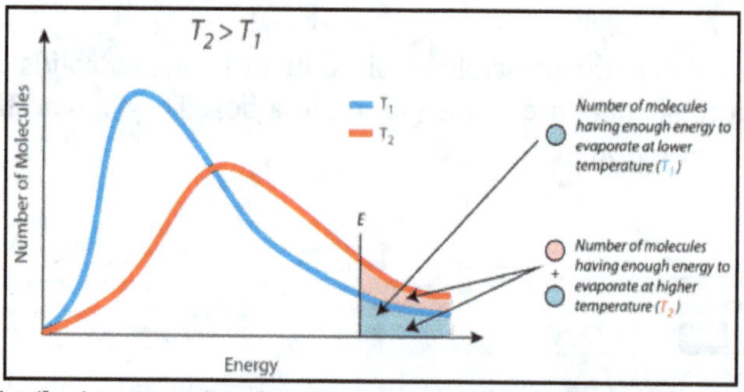

Figure: *Kinetic energy distribution curves for a liquid at two temperatures* T_1 and T_2. The shaded area is the molecules with enough kinetic energy to escape the liquid and become vapor.

Simple Distillation

Simple distillation is a procedure by which two liquids with different boiling points can be separated. Simple distillation (the procedure outlined below) can be used effectively to separate liquids that have at least fifty degrees difference in their boiling points. As the liquid being distilled is heated, the vapors that form will be richest in the component of the mixture that boils at the lowest temperature. Purified compounds will boil, and thus turn into vapors, over a relatively small temperature range (2 or 3 °C); by carefully watching the temperature in the distillation flask, it is possible to affect a reasonably good separation. As distillation progresses, the concentration of the lowest boiling component will steadily decrease. Eventually the temperature within the apparatus will begin to change; a pure compound is no longer being distilled. The temperature will continue to increase until the boiling point of the next-lowest-boiling compound is approached. When the temperature again stabilizes, another pure fraction of the distillate can be collected. This fraction of distillate will be primarily the compound that boils at the second lowest temperature. This process can be repeated until all the fractions of the original mixture have been separated.

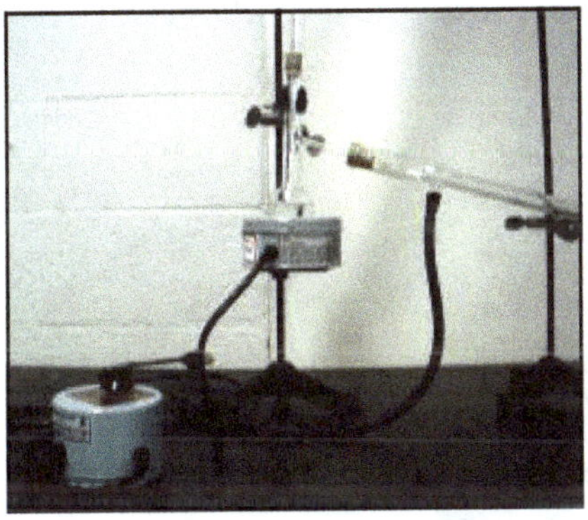

Figure: Distillation apparatus. A distillation flask with a thermometer is placed in a heating mantle and is connected to a condenser.

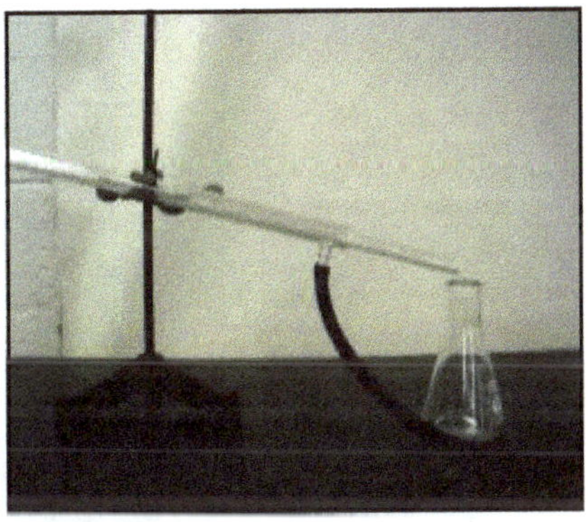

Figure: The tubes on the condenser are attached to a water source, with the water flowing in the low end and flowing out the high end of the condenser. The condensed vapor drips into the collection receiver.

Basic Procedure

1. Check the calibration of the thermometer that is to be used. This can be accomplished by placing the thermometer in an ice bath of distilled water. After the thermometer has been allowed to reach thermal equilibrium, place it in a beaker of boiling distilled water and again allow it to reach thermal equilibrium. If the temperatures measured deviate from the expected values by more than two degrees, obtain a new thermometer and check its calibration.

2. Fill the distillation flask. The flask should be no more than two thirds full because there needs to be sufficient clearance above the surface of the liquid so that when boiling commences the liquid is not propelled into the condenser, compromising the purity of the distillate. Boiling chips should be placed in the distillation flask for two reasons: they will prevent superheating of the liquid being distilled and they will cause a more controlled boil, eliminating the possibility that the liquid in the distillation flask will bump into the condenser.

Figure: The thermometer is inserted in the distillation flask through a hole in the cork stopper. The arm of the flask is inserted through a hole in the stopper of the condenser. Make sure these stoppers are airtight, or the vapor will escape.

3. Heat the distillation flask slowly until the liquid begins to boil. Vapors will begin to rise through the neck of the distillation flask. As the vapors pass through the condenser, they will condense and drip into the collection receiver. An appropriate rate of distillation is approximately 20 drops per minute. Distillation must occur slowly enough that all the vapors condense to liquid in the condenser. Many organic compounds are flammable and if vapors pass through the condenser without condensing, they may ignite as they come in contact with the heat source.

Figure: The distillation flask being heated in a heating mantle.

Figure: The collection receiver. The vapors condense and drip from the condenser into the flask.

4. As the distillate begins to drop from the condenser, the temperature observed on the thermometer should be changing steadily. When the temperature stabilizes, use a new receiver to collect all the drops that form over a two to three degree range of temperature. As the temperature begins to rise again, switch to a third collection container to collect the distillate that now is formed. This process should be repeated; using a new receiver any time the temperature stabilizes or begins changing, until all of the distillate has been collected in discrete fractions.

Note: All fractions of the distillate should be saved until it is shown that the desired compound has been effectively separated by distillation.

5. Remove the heat source from the distillation flask before all of the liquid is vaporized. If all of the liquid is distilled away, there is a danger that peroxides, which can ignite or explode, may be present in the residue left behind. Also, when all of the liquid has evaporated, the temperature of the glass of the filtration flask will rise very rapidly, possibly igniting whatever vapors may still be present in the distillation flask.
    - Never distill to dryness. The residue left in the distillation flask may contain peroxides, which could ignite or explode after all the liquid has distilled away.
    - Make sure that all joints are secured very tightly. If any vapor escapes at the connection points, it may come into direct contact with the heat source and ignite.
    - Never heat a closed system, the increasing pressure will cause the glass to explode. If the distillation flask has a tapered neck, the thermometer may be placed in such a way as to block to flow of vapors up the neck of the flask; in effect creating a closed system; make sure that if using a tapered neck flask, the thermometer is not resting in the lowest portion of the neck.

Simple distillation is effective only when separating a volatile liquid from a nonvolatile substance or when separating two liquids that differ in boiling point by 50 degrees or more. If the liquids comprising the mixture that is being distilled have boiling points that are closer than 50 degrees to one another, the distillate collected will be richer in the more volatile compound but not to the degree necessary for complete separation of the individual compounds.

The basic idea behind fractional distillation is the same as simple distillation only the process is repeated many times. If simple distillation was performed on a mixture of liquids with similar volatilities, the resulting distillate would be more concentrated in the more volatile compound than the original mixture but it would still contain a significant amount of the higher boiling compound. If the distillate of this simple distillation was distilled again, the resulting distillate would again be even more concentrated in the lower boiling compound, but still a portion of the distillate would be the higher boiling compound. If this process is repeated several times, a fairly pure distillate will eventually result. This, however, would take a very long time. In fractional distillation, the vapors formed from the boiling mixture rise into the fractionating column where they condense on the column's packing. This condensation is tantamount to a single run of simple distillation; the condensate is more concentrated in the lower boiling compound than the mixture in the distillation flask. As vapors continue to rise through the column, the liquid that has condensed will revaporize. Each time this occurs the resulting vapors are more and more concentrated in the more volatile substances. The length of the fractionating column and the material it is packed with impact the number of times the vapors will recondense before passing into the condenser; the number of times the column will support this is referred to as the number of theoretical plates of the column.

Since the procedures of simple distillation are so similar to those involved in fractional distillation, the apparatus that are used in the procedures are also very similar. The only difference between the equipment used in fractional distillation and that used in simple distillation is that with fractional distillation, a packed fractionating column is attached to the top of the distillation flask and beneath the condenser. This provides the surface area on which rising vapors condense, and subsequently revaporize.

The fractionating column is used to supply a temperature gradient over which the distillation can occur. In an ideal situation, the temperature in the distillation flask would be equal to the boiling point of the mixture of liquids and the temperature at the top of the fractionating column would be equal to the boiling point of the lower boiling compound; all of the lower boiling compound would be distilled away before any of the higher boiling compound. In reality, fractions of the distillate must be collected because as the distillation proceeds, the concentration of the higher boiling compound in the distillate being collected steadily increases. Fractions of the distillate, which are collected over a small temperature range, will be essentially purified; several fractions should be collected as the temperature changes and these portions of the distillate should be distilled again to amplify the purification that has already occurred.

Fractional Distillation

Fractional distillation is a special type of distillation designed to separate a mixture of two or more liquids that have different boiling points. The process involves heating the mixture and partial condensation of the vapors along a column (fractionating column), which is set up such that components with lower boiling points pass through the column and are collected earlier than components with higher boiling points. Generally, this method is used when the component parts boil at less than 25 °C from each other under a pressure of one atmosphere (atm).

The separation technology of fractional distillation is useful in both research and industrial settings. In industry, it is commonly used in petroleum refineries, chemical plants, and natural gas processing plants. Likewise, if a sample of air is liquefied, it can be separated into its components to produce liquid oxygen, liquid nitrogen, and argon. Also, chlorosilanes can be distilled to produce high-purity silicon for use as a semiconductor.

Fractional Distillation in a Laboratory

Apparatus

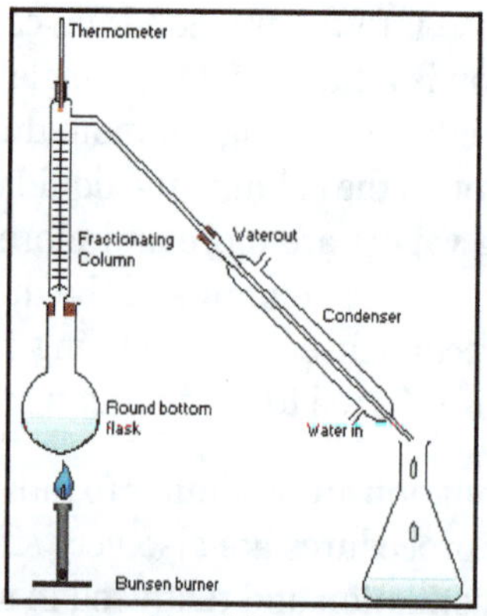

An example of a fractional distillation apparatus used in a laboratory. The diagram shows the use of a Liebig condenser and a conical flask as a receiving flask. Here, the distillation head and fractionating column are combined in one piece.

A laboratory setup for fractional distillation often involves use of the following pieces of apparatus, put together as shown in the image on the right:

- A heat source (such as a hot plate with a bath).
- A distilling flask (typically a round-bottom flask).
- A receiving flask (often a round-bottom flask or conical flask).
- A fractionating column (vigreux column).
- A distillation head.
- A thermometer and adapter if needed.
- A condenser (liebig condenser, graham condenser, or allihn condenser).
- A vacuum adapter (for distillations under reduced pressure).
- Boiling chips (also known as anti-bumping granules).
- Rubber bungs, unless laboratory glassware with ground glass joints is used, such as a quickfit apparatus.

Procedure

The apparatus is assembled as in the diagram (which represents a batch apparatus, as opposed to a continuous apparatus). The mixture is put into a round-bottom flask along with a few anti-bumping granules, and the fractionating column is fitted over the mouth of the flask. As the mixture boils, vapor rises up the column. The vapor condenses on the glass platforms, known as trays, inside the column, and runs back down into the liquid below. This is known as "refluxing" the distillate.

Only the most volatile fraction of the vapors remains in gaseous form all the way to the top of the column. This fraction passes into the condenser, which cools it down until it liquefies, and this liquid is collected in the receiving flask.

The efficiency in terms of the amount of heating and time required to get fractionation can be improved by insulating the outside of the column with an insulator such as wool, aluminum foil, or (preferably) a vacuum jacket. The hottest tray is at the bottom and the coolest is at the top. At steady state conditions, the vapor and liquid on each tray are at equilibrium. The fractionation is more thorough with the addition of more trays (up to certain practical limitations).

In laboratory distillation, any of several types of condensers may be used. The Liebig condenser is simply a straight tube within a water jacket, and is the simplest (and relatively least expensive) form of condenser. The Graham condenser is a spiral tube within a water jacket. The Allihn condenser has a series of large and small constrictions on the inside tube, each increasing the surface area upon which the vapor constituents may condense.

In alternate set-ups, a "cow" or "pig" adapter may be used, connected to three or four receiving flasks. By turning the "cow" or "pig," the distillates can be channeled into the appropriate receiver. A Perkin triangle may also be used to collect distillation fractions, without requiring a "cow" or "pig" adapter. A Perkin triangle is most often used when the distillates are air-sensitive, or when the fractions distill and are collected under reduced pressure, but it can be used for simple as well as fractional distillations.

Vacuum distillation systems operate at reduced pressure, thereby lowering the boiling point of the materials.

Example

Consider the distillation of a mixture of water and ethanol. Ethanol boils at 78.5 °C, and water boils at 100 °C. On that basis, one should be able to separate the two components by fractional distillation. However, a mixture of 96 percent ethanol and four percent water boils at 78.2 °C, being more volatile than pure ethanol. Such a mixture is called an azeotrope. When the mixture is gently heated, the azeotrope (being the most volatile component) concentrates to a greater degree in the vapor and separates from the rest of the liquid first. Thus, fractional distillation of a mixture of water and ethanol produces 96 percent ethanol. Once all the ethanol has boiled out of the mixture, the thermometer shows a sharp rise in temperature.

Industrial Distillation

Typical industrial fractional distillation columns.

Fractional distillation is the most common form of separation technology used in petroleum refineries, petrochemical and chemical plants, and natural gas processing plants. For example, this technique is used in oil refineries to separate crude oil into useful substances (or fractions) consisting of different hydrocarbons with differing boiling points. The crude oil fractions with higher boiling points consist of larger molecules (with more carbon atoms and higher molecular weights), are darker in color, are more viscous, and are more difficult to ignite and to burn.

Most often, new feed is continuously added to the distillation column, and products are continuously removed. Unless the process is disturbed due to changes in feed, heat, ambient temperature, or condensing, the amount of feed being added and the amount of product being removed are normally equal. This is known as continuous, steady-state fractional distillation.

Industrial distillation is typically performed in large, vertical cylindrical columns known as "distillation or fractionation towers" or "distillation columns." Their diameters range from about 65 centimeters to six meters, and their heights range from about six meters to 60 meters or more. The distillation towers have liquid outlets at intervals up the column, allowing for the withdrawal of different *fractions* or products with different boiling points or boiling ranges. The "lightest" products (those with the lowest boiling point) exit from the top of the columns and the "heaviest" products (those with the highest boiling point) exit from the bottom of the column.

Fractional distillation is also used for the separation of (liquefied) air into its components, producing liquid oxygen, liquid nitrogen, and high purity argon. Distillation of chlorosilanes enables the production of high-purity silicon for use as a semiconductor.

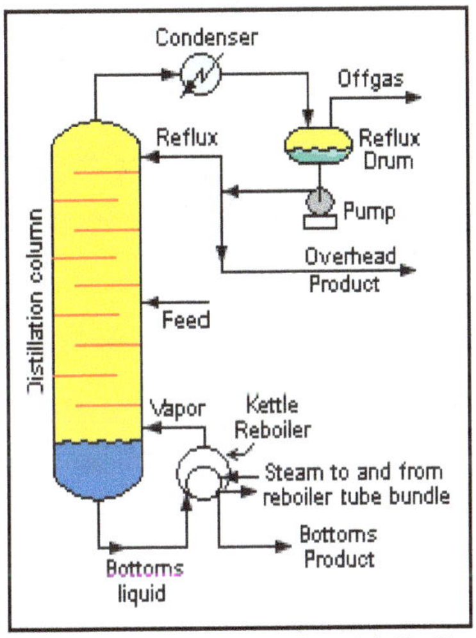

Diagram of a typical, large-scale industrial distillation tower.

Large-scale industrial towers use reflux to achieve a more complete separation of products. Reflux refers to the portion of the condensed overhead liquid product from a fractionation tower that is returned to the upper part of the tower as shown in the schematic diagram on the right. Inside the tower, the reflux liquid flowing downward provides the cooling needed to condense the vapors flowing upward, thereby increasing the effectiveness of the distillation tower. The more reflux is provided for a given number of theoretical plates, the better the tower's ability to separate lower boiling materials from higher boiling materials. Alternatively, the more reflux provided for a given desired separation, the fewer theoretical plates are required.

In industrial uses, sometimes a packing material is used instead of trays within the column, especially when operating under reduced pressures. This packing material can either be random dumped packing (one to three inches wide) such as Raschig rings or structured sheet metal. Liquids tend to wet the surface of the packing and the vapors pass across this wetted surface, where mass transfer takes place. Unlike conventional tray distillation, in which every tray represents a separate point of vapor liquid equilibrium, the vapor liquid equilibrium curve in a packed column is continuous. However, when modeling packed columns, it is useful to compute a number of "theoretical plates" to denote the separation efficiency of the packed column with respect to more traditional trays. Differently shaped packings have different surface areas and void space between packings. Both of these factors affect packing performance.

Design of Industrial Distillation Columns

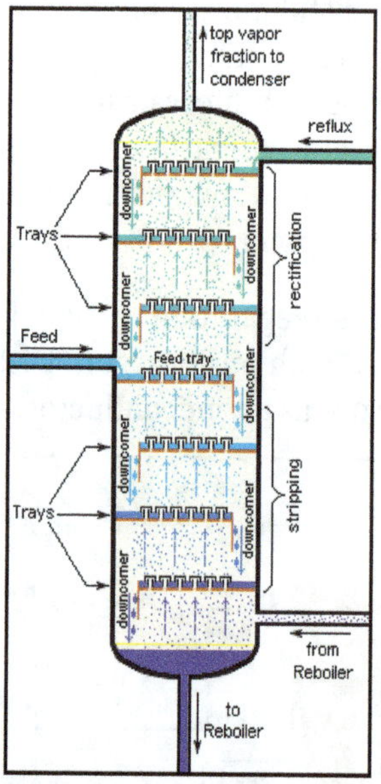

Chemical engineering schematic of typical bubble-cap trays in a distillation tower.

Design and operation of a distillation column depends on the feed and desired products. Given a simple, binary component feed, analytical methods such as the McCabe-Thiele method or the Fenske equation can be used. For a multi-component feed, simulation models are used both for design and operation.

Moreover, the efficiencies of the vapor-liquid contact devices (referred to as plates or trays) used in distillation columns, as seen in Figure two, are typically lower than that of a theoretical 100 percent efficient equilibrium stage. Hence, a distillation column needs more plates than the number of theoretical vapor-liquid equilibrium stages.

Centrifugation

Centrifugation separates on the basis of the particle size and density difference between the liquid and solid phases. Sedimentation of material in a centrifugal field may be described by:

$$V = \frac{d^2(\rho_s - \rho_I)\omega^2 r F_s}{18\eta\theta}$$

where v is the rate of sedimentation, d is the particle diameter, ρ_s is the particle density, ρ_I is the solution density, w is the angular velocity in radians s^{-1}, r is the radius of rotation, h is the dynamic viscosity, F_s is a correction factor for particle interaction during hindered settling and q is a shape factor (=1 for spherical particles). F_s depends on the volume fraction of the solids present; approximately equalling 1, 0.5, 0.1 and 0.05 for 1%, 3%, 12% and 20% solids volume fraction respectively. Only material which reaches a surface during the flow through continuous centrifuges will be removed from the centrifuge feedstock, the efficiency depending on the residence time within the centrifuge and the distance necessary for sedimentation (D). This residence time will equal the volumetric throughput (f) divided by the volume of the centrifuge (V). The maximum throughput of a centrifuge for efficient use is given by:

$$\phi = \frac{d^2\left(\rho_s - \rho_I\right)\omega^2 r V F_s}{18\eta\theta D}$$

The efficiency of the process is seen to depend on the solids volume fraction, the effective clarifying surface (V/D) and the acceleration factor (w^2r/g, where g is the gravitational constant, 981 cm s^{-2}; a rotor of radius 25 cm spinning at 1 rev s^{-1} has an acceleration factor of approximately 1 G). Low acceleration factors of about 1 500 g may be used for harvesting cells whereas much higher acceleration factors are needed to collect enzyme efficiently. The product of these factors (w^2rV/gD) is called the sigma factor (S) and is used to compare centrifuges and to assist scale-up.

Laboratory centrifuges using tubes in swing-out or angle head rotors have high angular velocity (w) and radius of rotation (r) but small capacity (V) and substantial sedimentation distance (D). This type of design cannot be scaled-up safely, primarily because the mechanical stress on the centrifuge head increases with the square of the radius, which must increase with increasing capacity.

For large-scale use, continuous centrifuges of various types are employed. These allow the continuous addition of feedstock, the continuous removal of supernatant and the discontinuous, semicontinuous or continuous removal of solids. Where discontinuous or semicontinuous removal of precipitate occurs, the precipitate is flushed out by automatic discharge systems which cause its dilution with water or medium and may be a problem if the precipitate is required for further treatment. Centrifugation is the generally preferred method for the collection of enzyme-containing solids as it does not present a great hazard to most enzymes so long as foam production, with consequent enzymic inactivation, is minimised.

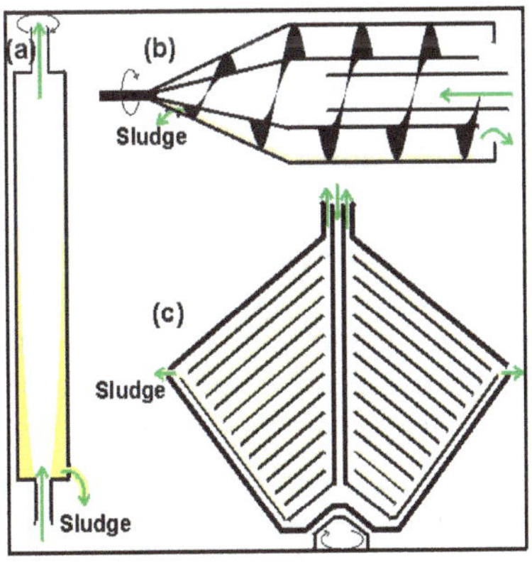

In figure basic designs of industrial centrifuges, showing the flow of material within the bowls. Motor drives, cooling jackets and sludge collection vessels are not shown. (a) Tubular bowl centrifuge. This is generally operated vertically, the tubular rotor providing a long flow path enabling clarification. The sludge collects and must be removed. (b) Continuous scroll centrifuge. This is operated horizontally. The helical screw scrolls the solids along the bowl surface and out of the liquid; the sludge being dewatered before discharge. The clarified liquor overflows over an adjustable weir at the other end of the bowl. The screw conveyer rotates at a slightly different speed to the bowl. (c) Continuous multichamber disc-stack centrifuge. The bowl contains a number of parallel discs providing a large clarifying surface with a small sedimentation distance. The sludge is removed through a valve.

Centrifuges

Centrifuge is any device that applies a sustained centrifugal force—that is, a force due to rotation. Effectively, the centrifuge substitutes a similar, stronger, force for that of gravity. Every centrifuge contains a spinning vessel; there are many configurations, depending on use. A perforated rotating drum in a laundry that throws off excess water from clothes, for example, is a type of centrifuge. A similar type is used in industry to separate fluids from solid matter after crushing.

As enunciated by Sir Isaac Newton in his first law of motion, a freely moving body (such as a ball) tends to travel in a straight line, and, if directed along a curved path by some restraining force (such as would result were a hand-held string tied to it), it will exert a force against the directing or restraining force in its continual effort to fly off onto a straight tangential course. It is a familiar observation that an object revolving in a circle exerts a force away from the center of rotation. This force, which is the outward pull of the ball on its string, is the centrifugal force. Also, there is general appreciation of the fact that the amount of this force can be increased by increasing either the angular velocity of rotation, the mass of the object, or the radius of the circle through which the object moves. Perhaps not so generally appreciated is the fact that, whereas the centrifugal force is directly proportional to the radius and to the mass, it is proportional to the square of the angular velocity. For example, doubling the mass of the rotating object will increase the centrifugal force by a factor of 2, but doubling the number of revolutions per minute (rpm) will increase the centrifugal force by a factor of 4 (equals 2 times 2); similarly, increasing the speed by a factor of 10 will increase the force by a factor of 100 (equals 10 times 10). Centrifugal force is expressed by the basic relation $F = mv^2 / R = 4\pi^2 mn^2 R$; F is the centrifugal force, m the mass, R the radius, v the speed, and n the number of revolutions per second.

The centrifugal force is often compared directly with the weight (pull of gravity) of the object, and the amount of force is stated as so many "times gravity" or so many "*g*." Through the use of special research apparatus, forces greater than 5,000,000 times gravity have been produced by spinning small metal rotors of about pea size at speeds exceeding 1,000,000 revolutions per minute.

The rotating element of a centrifuge is usually driven about a fixed axis by an electric motor, or by an air turbine in some high-speed machines, and is known as a rotor, bowl, or drum. For the minimizing of vibration and strain on the shaft and bearings, it is essential that a loaded rotor be well balanced—i.e., that its total mass be so distributed about the axis of rotation that the resultant of all the elemental forces is zero. If the bearings are suited to high speeds and if ample power is available to overcome the frictional resistance of the bearings, the only limitation to the speed of a well-balanced rotor is the strength against rupture of the material from which it is made.

For example, a rotor with a 15-cm (6-inch) diameter used in certain biological studies and designed especially for high speeds has a limiting speed for routine operation of about 60,000 revolutions per minute. In a rotor of given design, the maximum angular velocity obtainable before rupture is to a close approximation inversely proportional to the rotor's diameter. Thus, a small rotor having only one-half the diameter of a larger one can be as safely rotated at twice the angular velocity and with the production at the periphery of twice the centrifugal force.

The widest use of centrifuges is for the concentration and purification of materials in suspension or dissolved in fluids. Suspended particles denser than the suspending liquid tend to migrate toward the periphery, while those less dense move toward the center. The rapidity with which the migration proceeds is dependent on the intensity of the centrifugal field, the difference between the density of the particle and that of the suspending liquid, the viscosity of the liquid, the size and shape of the particle, and to some extent the concentration of the particles and the degree to which they are electrically charged. The net motivating force exerted on the particle is the difference between the centrifugal field acting on it and the opposing buoyancy of the liquid. All other things being equal for two particles, one with a diameter 10 times that of the other will require only $^1/_{100}$ as much average centrifugal field to move a given distance in a given time as the smaller.

From the foregoing discussion, it is clear that a practically complete separation of the suspending medium and the suspended particles can be produced if the centrifugation is allowed to continue until all particles have collected against the outer wall of the spinning vessel or centrifuge. It should also be noted that a partial separation of two groups of suspended particles of different size can be effected by allowing centrifugation to continue only long enough for all of the larger particles to be completely packed into the sediment, since then many of the small particles will still be suspended in the fluid. If separation of the larger as well as the smaller particles is desired, the surface fluid can be drawn off and the sediment re-suspended in some suitable liquid and subsequently centrifuged again to effect further separation.

Centrifuges may be classified in three general categories depending on whether the spinning centrifuge bowl that contains the material to be separated has a solid wall, a perforated wall, or some combination of the two. Also, they may be characterized according to whether the material is treated in a continuous flow process, a batch process, or a combination of the above processes.

Bottle Centrifuges

A bottle centrifuge is a batch-type separator that is primarily used for research, testing, or control. The separation takes place in test tube or "bottle-type" containers, which are symmetrically mounted on a vertical shaft. The shaft of a bottle centrifuge is usually driven by an electric motor, gas turbine, or a hand-driven gear train located above or below the rotor. In most cases, the bottles are supported by high-strength metal containers so that their axis is perpendicular to the axis of rotation. The sedimentation occurs in a radial direction, and in some bottle centrifuges the test tubes or bottles are inclined at an angle of about 37° to the axis of rotation in order to reduce the distance that the material must settle.

Bottle centrifuges are standard equipment for most biological, chemical, or medical laboratories. They are used to separate solid materials in suspension or to clarify liquids when precipitation will not take place in a reasonable time in the gravitational field g. In most commercial bottle centrifuges the centrifugal field may be varied from a few g up to tens of thousands. Commercial uses of the bottle centrifuge include tests for the butterfat content of milk, determination of the sediment in crude mineral and vegetable oils, and clinical tests of various kinds.

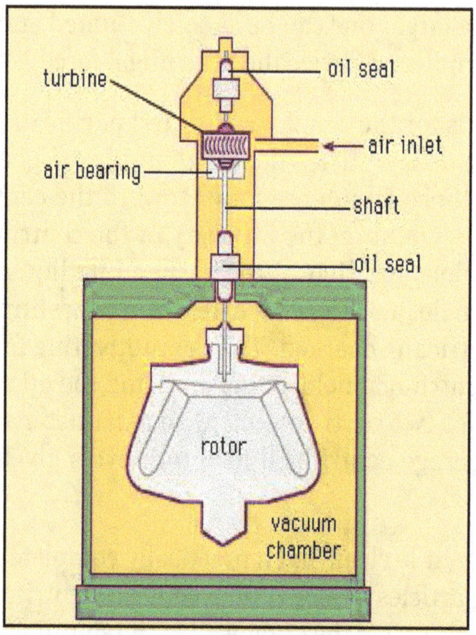

Tubular Centrifuges

The tubular centrifuge is used primarily for the continuous separation of liquids from liquids or of very fine particles from liquids, although in some cases it is employed as a batch-type centrifuge. In general, it is used when higher centrifugal fields are required for separation. The rotating bowl of a tubular centrifuge consists of a long hollow tube (length many times its diameter) as shown schematically in figure. For continuous separation the feed or material to be centrifuged enters at one end near the axis and is removed in two streams containing the separated material. In many cases the separation is not complete, and the separated fractions must be passed through the machine several times. Many different designs for the internal structure of the tube are employed, but, in general, radial vanes are used to bring the feed material up to speed and to slow down the separated streams before they are discharged. The centrifuge is driven by a high-speed motor or an air or steam turbine. The sedimentation takes place as the fluid flows from one end of the tube to the other. When the heavy material consists of very fine particles or molecules and the concentration is very low, the solid material is usually allowed to deposit on the wall. In this case the machine is operated as a batch centrifuge.

The tubular centrifuge is finding an increasing number of applications because of the high centrifugal fields that may be used ($10^5 g$ near the periphery in some cases). A few typical uses are as follows:

1. The purification of vaccines (un-centrifuged vaccines contain a large amount of nonessential and harmful material).
2. Purification of lubricating and industrial oils.
3. Clarification and purification of food products such as essential oils, extracts, and fruit juices.
4. Separation of immiscible liquids that cannot be separated by gravity.

Disk-Type Centrifuges

The disk-type centrifuge consists of a vertical stack of thin disks in the shape of cones. The sedimentation takes place in the radial direction in the space between adjacentcones. This greatly reduces the settling distance and hence increases the rate at which the material is separated. The angle of the cones is designed so that upon reaching the inside surface of the cone the heavier material slides down along its surface in a manner that is similar to that of the 37° fixed-angle bottle centrifuge.

The disk-type centrifuge usually operates continuously. The material to be processed enters in one stream and is separated into two purified streams. These centrifuges are used primarily for the separation of liquids in which the solid or immiscible components occur in relatively low concentrations. The familiar cream separator, widely used in the dairy industry and on farms for separating cream from milk, is a typical example of this type of centrifuge. They also are used for the purification of fuel oil, the reclamation of used motor oil, and the removal of soap stock in the refining of vegetable oils.

Basket Centrifuges

Basket centrifuges are often called centrifugal filters or clarifiers. They have a perforated wall and cylindrical tubular rotor. In many cases the outer wall of a basket centrifuge consists of a fine mesh screen or a series of screens with the finer mesh screens supported by the heavier coarse screen, which in turn is supported by the bowl. The liquid passes through the screen, and the particles too large to pass through the screen are deposited. The basket centrifuge is employed in the manufacture of cane sugar, in the home and in laundries for the rapid drying of clothes, and in the washing and drying of many kinds of crystals and fibrous materials, etc.

Vacuum-Type Centrifuges

In the centrifuges described above, the rotor spins in air or some other gas at atmospheric pressure. The gaseous friction on a spinning rotor increases at a relatively high rate so that the power required to drive the rotor also increases rapidly. As a result, the temperature of the rotor rises drastically, sometimes exceeding the boiling point of water. As the rotor surface near the periphery moves faster than near the axis, a thermal gradient or variation in temperature through the rotor wall is established along the radius with the periphery at a higher temperature than the axis. These small radial temperature gradients produce convection within the centrifuge, and these convection currents can cause remixing and disturb sedimentation.

The heat buildup and convection problems caused within a centrifuge by air resistance can be avoided by spinning the rotor within an evacuated chamber. The elimination of air resistance also makes possible the attainment of high rotational speeds with relatively little expenditure of energy. Many vacuum-type centrifuges are ultracentrifuges; i.e., they operate at speeds of more than about 20,000 revolutions per minute. figure shows a schematic diagram of an early vacuum-type ultracentrifuge. The centrifuge rotor located inside the vacuum chamber is connected to the air-supported, air-driven turbine by a vertical, small-diameter, flexible steel shaft.

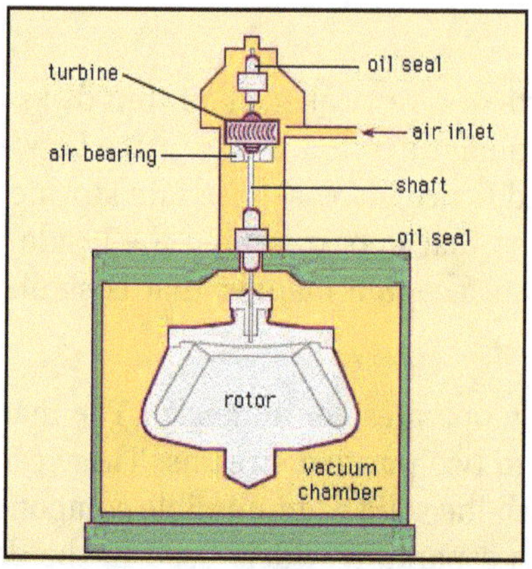

Figure: Air-driven, vacuum-type ultracentrifuge.

The rotor of a typical vacuum-type ultracentrifuge is 18 cm (7 inches) in diameter and carries 300 ml (10 ounces) of liquid in a centrifugal field of more than 300,000*g*. Practically all substances of importance in medicine and biology and all other substances with molecular weights of 50 daltons (one dalton is 1.66×10^{-24} grams) or more are easily purified in this type of bottle centrifuge. The rotor of a vacuum-type ultracentrifuge can be replaced by one with sector-shaped cells and transparent windows so that the progress of the sedimentation can be optically measured and photographed. This method was first used by T. Svedberg and J.B. Nichols in 1923 and was widely applied thereafter to determine the sedimentation rates and sizes of many submicroscopic particles, particularly protein molecules and viruses.

The vacuum-type centrifuge may be used for the determination of the molecular weights of practically all substances in solution. In modern commercial vacuum-type centrifuges the air drive and support have been replaced by the more efficient and convenient electric motor drive, and the entire machine has been redesigned and made almost automatic in its operation. The present commercial vacuum-type ultracentrifuge has become an indispensable tool in laboratories where it is necessary to purify substances of importance in biochemistry, biophysics, biology, medicine, and the pharmaceutical industry.

The ultracentrifuge can be used in two principal ways for determining the molecular weights of various proteins. The first consists in carrying out the sedimentation in a centrifugal field high enough to produce a relatively sharp sedimentation boundary—i.e., the boundary between the sedimenting molecules and the pure solvent. The rate at which this boundary moves out along the radius toward the periphery is then measured and the value of the molecular weight is calculated. This is called the rate of sedimentation method. The second method consists in centrifuging the material until equilibrium is established in the centrifuge cell—i.e., until the rate at which the material settles out is balanced by back diffusion. If the concentration in the cell is then determined at various radial distances, the value of the molecular weight can be calculated.

Vacuum-type tubular centrifuges are used to purify many biological materials that cannot easily be separated in other ways. They have been employed both as continuous-flow and as density-gradient centrifuges. The density-gradient centrifuge consists in setting up a radial density variation or gradient in the tubular centrifuge with slowly sedimenting nonreactive smaller molecules such as sucrose or calcium chloride. If, then, the density of the substance to be purified falls within the range of the artificial density gradient, it will collect in a thin cylindrical surface at a definite radius. If more than one substance is in the solution, each of the substances will collect at a radius determined by its particle density.

Another important use of the vacuum-type centrifuge is gas separation. When a gas is subjected to a centrifugal field, a radial pressure gradient is immediately established. Consequently, a mixture of any two gases with different molecular weights may be separated in a centrifuge with the lighter gas being concentrated on the axis. In 1919, after it was pointed out that it should be possible to separate the isotopes of an element by centrifuging, a number of attempts were made to obtain separation but were all unsuccessful, probably due to convection and remixing in the centrifuge. In 1937 the isotopes of chlorine were separated with a vacuum-type ultracentrifuge. An evaporative centrifuge method was used in which the material to be separated is admitted to the rotor and condensed on the periphery with the rotor stationary. The rotor is then driven to operating speed and the lighter material pumped out through the hollow shaft while the heavier material remains in the centrifuge to be collected later. The centrifuge used in gas separation should be spun as rapidly as possible and should be as long as possible. The centrifuge method is suited to the separation of the heavier isotopes as well as the lighter ones, because it depends on the differences in the masses rather than on their absolute values.

Since the mid-1940s the technique of gaseous centrifuging has been further developed and extended. Workers in Germany and in the Netherlands have had considerable success with the method. A remarkably simple vacuum-type gas centrifuge that is especially adapted to uraniumisotope separation has been devised. During the 1970s a centrifuge plant was constructed in Europe for the purpose of commercially producing reactor-grade uranium-235 for use in nuclear power plants.

Magnetic Separation

The basic principle behind magnetic separations is remarkably simple and remains unchanged from these early examples. It relies on the simple fact that materials with differing magnetic moments experience different forces in the presence of magnetic field gradients; thus, an externally applied field can handpick out of physically similar mixtures those components with distinctive magnetic characteristics. The use of this principle is straightforward in mixtures where a magnetic component is known to exist; an intrinsically magnetic material can be separated using electrically powered electromagnets or strong permanent magnets. The process is generally binary and results in a magnetically rich retentate (usually a solid) and the residual non-magnetic solvent.

The 1950s were a time of great expansion for the field of magnetic separations as the introduction of high gradient magnetic separation (HGMS) systems permitted faster and more general magnetic separation processes. HGMS works through the application of large static fields ($\sim 1\ T$) to columns containing ferrous matrices such as steel wool; these irregular surfaces give rise to magnetic gradients as high as 10^4 T/m which generate forces large enough to capture even weakly magnetic particles in a flow stream. This new technology meant that magnetic separations could be applied more universally to separation problems where particulate matter was not strongly magnetic. The use of HGMS in water treatment, for example, provided a way to clarify water with high sediment loads, provided that the sediment has some weakly magnetic character. A larger market was found in the steel industry where HGMS separators became critical components for purifying high grade, low iron content steel.

More recently, separations using external magnetic fields have become commonplace in biotechnology where they are used for both protein purification as well as flow cytometry. In these contexts, separators apply relatively small field gradients (500 T/m) to fixed volumes of solution with the aim of removing a valuable biological components. Such an approach to separations is more as a `batch process', and it works well for biomedical research laboratories where solution volumes are small and separation speed is not a major issue. However, the small scale and time consuming nature of biomagnetic separators precludes their application in biological manufacturing processes. In addition, the separators themselves are standalone configurations of permanent magnets which require little upkeep or even power. This separator simplicity is offset by the need for magnetic beads that are coated so as to bind to particular biological components. As a result, the technique is not that general and conventional biomedical researchers can only use magnetic separations if a commercial magnetic bead source is available for their targets. Commercial sources for magnetic beads have grown substantially in the past decade, however, and a wide variety of proteins, cells and other biomacromolecules may be selectively removed using these methods.

Magnetic Separations: Their Unique Position among Separation Technologies

The problem of separating materials, whether they are specialized chemicals, high purity steel or valuable pharmaceuticals, from product streams is a near universal one for any manufacturing process. When speed is not an issue and the materials of interest are solids or flocculated products, sedimentation or centrifugation is routinely employed. For faster processing, filtration is a well-established method for removing waste or concentrating product. Given these standard methods it is not always obvious that magnetic separations could or should be applied to a given problem. Certainly, if part of a mixture is intrinsically magnetic then magnetic removal is often the best solution. In these cases, magnetic separations generally offer higher throughput with greater specificity than equivalent centrifugation or filtration methods.

However, even in the absence of intrinsically magnetic components the use of designer magnetic beads—targeted to the product of interest—can make a magnetic separation feasible for virtually any system. Such processes offer very different kinds of tradeoffs in speed and selectivity as opposed to the more conventional approaches. A consideration of the unique advantages of moving materials with external fields, as opposed to other conventional alternatives, is detailed in table. For the purposes of this analysis we considered both the slower `batch' model for magnetic separations as well as the faster flow separators based on the high gradient magnetic columns. We also limit our comments to separation processes for materials in liquids, a case which captures the majority of current application in this area.

An important feature of a magnetic separation is that the product stream meets with virtually no flow resistance as it moves through a separator; this is in stark contrast to filters which use a solid phase as the basis for the separation process. Filtration is particularly challenging when submicron particles or biomolecules are the target products; this requires filter pore sizes well under a micron (ultrafiltration) or even a submicron (nanofiltration). The flow resistance that this solid phase introduces to the system is significant, and industrial membrane filters often must operate with hundreds of pounds of solution pressure (many times the flow of a fire hydrant) to maintain flow rates of several milliliters per minute. Such units require not just energy-intensive pumping stations, but also expensive and high performance fittings and valves. Filters also can become fouled with a wide variety of contaminants requiring backwashing procedures and ultimately replacement.

Conventional magnetic separators face few of these problems. In high-gradient flow systems, the columns are loosely packed with steel wools that offer virtually no resistance to flow. Reasonable operating pressures and conventional pumps are suitable for these systems, though the electromagnets that are needed for very high gradient separations do require significant energy. For separations in which speed is not an issue, then the batch magnetic separators are interesting options. Because they usually generate the fields with permanent magnets, they can operate without any external power; with the appropriate beads, such systems can remove a greater variety of substances than those accessed through sedimentation. This makes them particularly attractive for small-volume separations in low infrastructure settings.

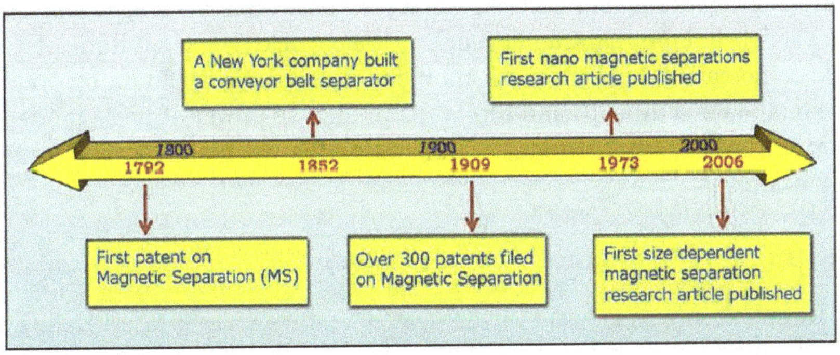

Figure: The evolution of magnetic separation technology.

Table: Industrial applications of magnetic separations.

Application area	Objective
Chemical and related industries	Tramp metal removal from machinery to avoid wear and malfunction
Food industries	
Coal processing plants	
Metals production and recycling industries	
Raw materials processing plants, such as cement, glass, semiconductors	Removal of iron and derivatives contamination
Mineral post processing industries	Separation and enrichment of magnetic ores (magnetite, hematite, ilminite, etc.)

While magnetic separations are in principle excellent choices for many problems, their widespread application has been limited by the complexity required in separator design and bead technology. In the case of flow separations, cumbersome and expensive electro- magnets are used to create external fields in excess of a Tesla; the requirement for high gradients also limits column diameters and thus reduces throughput. In the batch separations favored in biotechnology, permanent magnets in a multipole configuration provide the relatively low gradients needed and their set-up and use is easy and immediate. The trade-off here is that these low fields produce only very small magnetic forces on particles. Since the magnetic force is proportional to the particle volume, the small field gradients are generally offset by the use of larger magnetic beads. This lowers the available surface area for biomolecular recognition and adsorption, and has precluded application of this technology to biological separations on the larger scale. Our group and others are working on ways to better design nano scale magnetic beads so as to overcome these and other limitations. This materials design process must be guided by a more quantitative and formal description of the process of magnetic separations.

Modeling Magnetic Separations

The magnetic separation process is amenable to simple modeling and such treatments date back to the 1950s; most analyses assume that the materials being removed act independently and that macroscopic models for fluid dynamics are appropriate. These constraints make it straightforward to describe the many forces acting on particles in a flow, such as the magnetic force, Brownian motion, gravitation, and friction. What results is a universal equation that relates the net force on a particle to input parameters such as the particle size, magnitude of the field gradient, frictional coefficient, and the magnetic moment of the particle. Obtaining precise values for these experimental data is difficult and limits the predictive power of these models. Still, these treatments provide an essential foundation for improving the materials and systems used for magnetic separations.

Simple Treatments for Magnetic Separations

Magnetic separation occurs as a result of the diverse and competing forces acting on a particle during flow in the presence of an external field. These include a hydrodynamic drag force due to the flow velocity; a magnetic force due to the gradient of the applied magnetic field; gravitational forces for large and dense particles; and finally a diffusive force which, arises from the intrinsic Brownian motion of particles. While the first three are deterministic forces that control a particle's trajectory, the latter is a randomizing force that must be exceeded if an ensemble of particles is to experience net movement during some time interval. Put in other terms, if the energy available to move a particle does not exceed its thermal energy due to Brownian motion then it is not possible to set up a concentration gradient.

Perhaps the most important factor to include in models of magnetic separations is the role of inter-particle forces such as Helmhotz double layer interaction, dipole–dipole interaction and Van Der Waals attraction. Thus, the magnetic, dipole–dipole interaction and Van Der Waals forces aid the process of separation, whereas, diffusion, double layer interaction and drag force act against the separation. Also, the particles are found to reversibly aggregate in presence of high field gradients, which would increase the net magnetic moment to a magnitude much stronger than the competing forces.

Magnetic Separations use in Column Formats: Examples

Magnetic separators used on the industrial scale are overwhelmingly high gradient systems which function as a column in continuous flow operations.

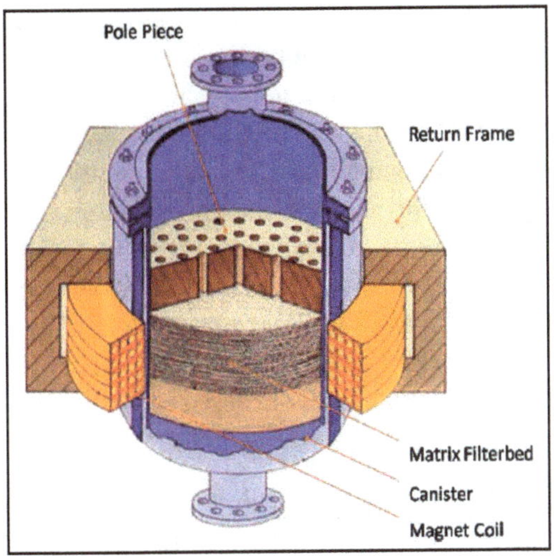

Figure: Metso® high gradient magnetic separators (HGMS) are designed to recover weakly magnetic material from non-magnetic matter and can be used for many applications including the processing of clays, iron ores, rare earths and industrial minerals. In addition to the strongly magnetic minerals of Fe, Co, and Ni, a vast number of weakly magnetic minerals, which are not normally treatable by ordinary magnetic separators, may be processed by high gradient magnetic separators. Metso HGMS separators are able to remove even weakly paramagnetic materials (from Metso minerals).

The speed and efficiency of these systems have been exploited for decades and the early stages of the industry were marked by notable optimism: "Virtually every process in the chemical engineering industry is a potential application (for HGMS). Many previously unthinkable processes will now become practical, and many previously practical ones will become unthinkable. It has already happened in the kaolin industry, and is beginning to hap- pen elsewhere." While these systems are not as ubiquitous now as their inventors anticipated thirty years ago, their applications have expanded notably.

Kaolin (Clay) Decolorization

Kaolin (a.k.a. china clay) is a clay mixture primarily consisting of kaolinite ($Al_2O_2SiO_2.2H_2O$ or $Al_2Si_2O_5(OH)_4$) mineral. It is named from the Kaoling Hills of the city of Chingte chen where fine Chinese porcelains were produced; its resistance to acids and alkalis alike was greatly prized. Today, however, it's mostly used in the paper manufacturing industry where it plays dual role, as a filler between the pulp fibers and as a surface coating for a white glossy finish.

Natural kaolin has color as mined due to the iron containing micas, tourmaline, pyrite, anatase and rutile present in the material. To remove these impurities, kaolin can be magnetically cleaned with a continuous high gradient magnetic separator to produce highly white material suitable for paper or porcelain. Figure shows kaolin mineral before and after the decolorization process. Because of the resistance of kaolin impurities to other chemical cleaning methods, HGMS handles 75% of the world production of white porcelain and paper (Oder, 1976). A typical plant would have an HGMS with a filter diameter of about 2 m and capacity up to 20 t/h.

On average, generating 1 t of steel requires 151 t of water for cooling and cleaning purposes; the resulting wastewater is filled with many magnetic particulates and other iron-containing impurities. Those particles, especially when present in gas and hot water streams, cause significant problems in processing and must be removed. Conventional methods for cleaning steel mill waste and process waters include sedimentation, flocculation followed by sedimentation, and fixed bed filtration. Such approaches require either large areas for settling tanks and clarifiers or expensive and short- lived filter systems. Magnetic separation has emerged as an ideal solution for this industry and it has offered great time, space and cost savings. In a sample treatment at Kawasaki Steel Corporation of Japan, a 3 kOe field strength, 2.1 m diameter magnetic filter removes 80% of contaminants from the cooling wastewater of vacuum degassing process. Similar use in treatment of wastewater can also be found in power plants (both conventional and nuclear). For these cases, HGMS is used to remove ferromagnetic or paramagnetic particulates which extends the lifetime of cooling systems.

Magnetic separations can also be used to treat pollution. Fly ash from coal power plants is 18% iron oxide. Magnetic filtration has been applied to capture 15% of waste fly ash, thus providing a means for recycling. Estimates show that this can replace some of the magnetite used in industry. Figure shows an example of a ball mill separator used in these operations.

Enrichment of Ores–mineral Beneficiation

The treatment of ores with magnetic separation is carried out primarily to enrich iron-containing ores. Conventional chemical and settling methods are not suited for this purpose given the similar density and reactivity of transition metal minerals. The magnetic nature of iron species, however, is unique and thus a natural target for magnetic separations. Among the iron ores taconite is most often subjected to magnetic treatments. From a taconite ore (33% iron) Kelland (1973) was able to recover iron at 95% on a 5 cm/s flow rate. Today, Metso Minerals, Inc. (formerly Sala International AB) offers magnetic separators that can separate iron from ores with nearly 100% efficiency (depending on the particulate sizes, magnetic field and flow rate). Figure shows a successful continuous HGMS separator used for these purposes.

Magnetic separation of pyrite (FeS2) from coal for desulfurization is also a common process. The weakly magnetic nature of pyrite, however, requires that the raw ore be pre-treated thermally to convert the pyrite (FeS_2, Ms = 0.3 emu/g) to more strongly magnetic pyrrhotite (Fe_7S_8, Ms = 22 emu/g). Up to 91% removal of sulfur from coal can be achieved by microwave heating followed by a magnetic separation. Figure shows an industrial scale drum separator used in large scale applications for powders such as coal.

Food Industry

Strict food quality standards require the food products to be contaminant free, where mainly rare earth elements (REEs) constitute the majority. The food industry, therefore, has found magnetic separations to be an ideal method to remove REEs from food ingredients. Similar to the ore beneficiation or desulfurization, the target substances are weakly magnetic and require the high gradients of a magnetic field to be removed in a continuous food production line. Bunting Magnetics Co. offers magnetic metal separators and metal detectors for the quality of food and extended service life of the processing equipment, especially for cheese processing, chocolate plants, pet food processing, flour mills, spice plants, vegetable processing.

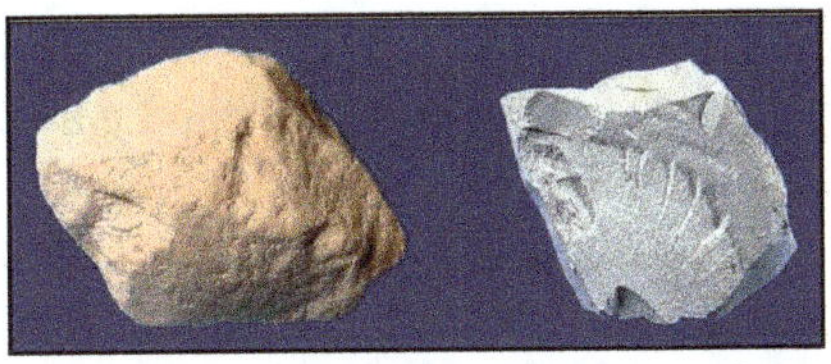

Figure: Kaolin decolorizes to white after magnetic separation.

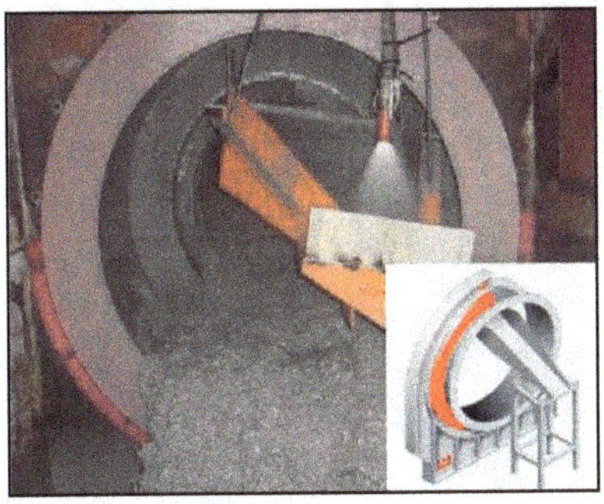

Figure: A ball mill separator. Separation of ball mill grinding ball segments from the discharge.

Removal of both ferrous and nonferrous tramp metals is achieved by their line of food safety products for the food processing industry. Figure shows case studies from Greenwood Magnetics Ltd., another company that produces magnets and assembly systems for cleaning REEs off of the food production lines.

Water Treatment and Metal Removal

With the new, lowered maximum permissible concentration for arsenic in drinking water (10 µg/l), effective since January 2006, techniques for better arsenic remediation without much desorption have gained more importance. Currently, coprecipitation, adsorption in fixed-bed filters, membrane filtration, anion exchange, electrocoagulation, and reverse osmosis are of methods of interest, however, cost efficiency and waste quantity requires further development that would aid in resolving the problem. Arsenic adsorption and desorption are heavily influenced by ad- sorbent particle size. Nanoscale magnetite (Fe_3O_4, 12 nm) can remove 200 times better than its commercial counterparts, which allows a significant cut in waste, instead of 1.4 kg of bulk iron oxide to remove arsenic (500 µg/l) from 50 l of solution, 15 g of nano magnetite can be used.

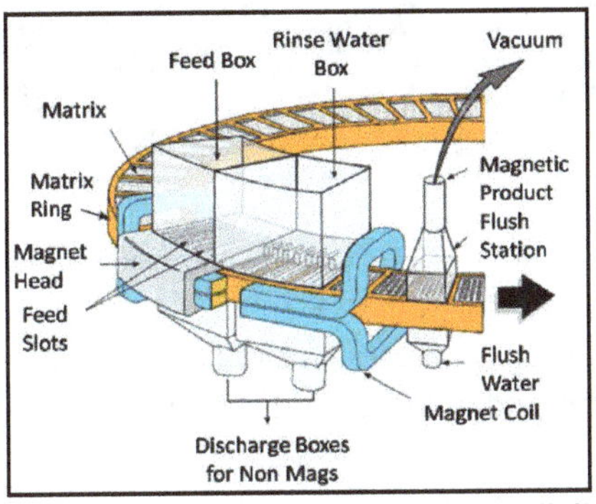

Figure: Continuous high gradient magnetic separation for many low susceptibility minerals that are associated with other minerals or have extra Fe in the crystals, and are hence often possible to separate.

Apart from surface area increment, an obvious gain while going down to nanoscale, available open sites with the proper chemistry (free Fe on the surface) can be accounted for this unexpected result. Size dependent magnetic properties bring controllability and along with mobility, a critical nanoscale advantage, provides unique applicability, if put in a system, in especially household uses where electricity is not readily available.

As early as 1970s Delatour (1973) and Delatour and Kolm (1975) treated water samples from the Charles River (Fe_3O_4 seeding, 5 ppm Al^{3+}) with a high flow velocity HGMS (V_0 = 136 mm/s, H_0 = 1 T) and reduced coliform bacteria from 2.2×105/l to 350/l, turbidity by 75%, color by 95%, and suspended solids by 78%. Later, Bitton and Mitchell removed 95% of the viruses from water by magnetic filtration following a 10 min of contact period with magnetite (added to be 250 ppm). The following years, Boliden Kemi AB reduced phosphorus of water supplies at least 87%. Also known as Sirofloc process, micron-sized magnetite is also used to remove color, turbidity, iron, and aluminum from water sources as an alternative to metal-ion coagulation. Recently, Denizli, and coworkers magnetically modified yeast cells for facile capture of mercury with fast biosorption rates (within 60 min) and efficiency (76.2 mg/g for Hg^{2+}).

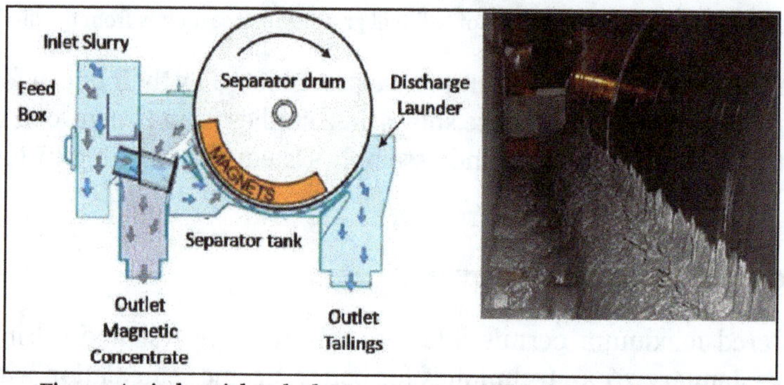

Figure: An industrial scale drum separator used in large scale applications.

(Left) A single row easy-clean grid box which contains high-density rare earth easy-clean magnetic tubes filters loose tea with a flow rate of 5 tph. (Middle) A water-jacketed pipeline magnet was manufactured to suit a 4 in pipeline pressure resistant up to 10 bar, nine high intensity rare earth magnets (11 500 G) filter liquid chocolate flowing at 300 l/min. The pressurized heated water-jacket maintains the temperature of the chocolate. (Right) The Bullet magnet that is used by a flour producer in which flour flows upwards through the 5 in pipe and any ferrous contamination is removed by the high intensity rare earth bullet magnet (8500 G min).

Table: Sources of contaminants in a steel production process.

Source of contaminant	Contaminants
Coke production	Non-magnetic particles, organics, and oils
Iron manufacturing	Magnetic particles and organics
Steel production	Magnetic particles
Cold finishing	Magnetic particles, oils, and acids
Hot formation	Magnetic particles and oils

Table: Arsenic removal by nano-sized magnetite.

Particle size (nm)	As (V) or As (III)	Residual as concentration (µg/l)	% Removal
12	As (III)	3.9	99.2
20	As (III)	45.3	90.9
300	As (III)	375.7	24.9
12	As (V)	7.8	98.4
20	As (V)	17.3	96.5
300	As (V)	354.1	29.2

References

- Sublimation-decanting-and-centrifuging: theartofseparationtechniques.blogspot.com, Retrieved 08, March 2020
- Applications-of-chromatography: chromatography.conferenceseries.com, Retrieved 19, June 2020
- Fractional-distillation: newworldencyclopedia.org, Retrieved 18, May 2020
- What-is-sublimation: studyread.com, Retrieved 17, February 2020
- Chromatography: explainthatstuff.com, Retrieved 05, July 2020
- Definition-of-decantation-604990: thoughtco.com, Retrieved 01, August 2020
- Liquid-chromatography: chemicool.com, Retrieved 21, January 2020

4

Fundamental Concepts of Chemical Engineering

There are various aspects which are vital to develop an understanding of chemical engineering. A few of these are transport phenomena, chemical thermodynamics, process simulation and chemical kinetics. This chapter closely examines these key concepts of chemical engineering to provide an extensive understanding of the subject.

Transport Phenomena

In engineering, physics and chemistry, the study of transport phenomena concerns the exchange of mass, energy, and momentum between observed and studied systems. While it draws from fields as diverse as continuum mechanics and thermodynamics, it places a heavy emphasis on the commonalities between the topics covered. Mass, momentum, and heat transport all share a very similar mathematical framework, and the parallels between them are exploited in the study of transport phenomena to draw deep mathematical connections that often provide very useful tools in the analysis of one field that are directly derived from the others. While it draws its theoretical foundation from principles in a number of fields, most of the fundamental transport theory is a restatement of basic conservation laws. The fundamental analyses in all three subfields of mass, heat, and momentum transfer are often grounded in the simple principle that the sum total of the quantities being studied must be conserved by the system and its environment. Thus, the different phenomena that lead to transport are each considered individually with the knowledge that the sum of their contributions must equal zero. This principle is useful for calculating many relevant quantities. For example, in fluid mechanics, a common use of transport analysis is to determine the velocity profile of a fluid flowing through a rigid volume.

Transport phenomena are ubiquitous throughout the engineering disciplines. Some of the most common examples of transport analysis in engineering are seen in the fields of process, chemical, biological, and mechanical engineering, but the subject is a fundamental component of the curriculum in all disciplines involved in any way with fluid mechanics, heat transfer, and mass transfer. It is now considered to be a part of the engineering discipline as much as thermodynamics, mechanics, and electromagnetism.

Transport phenomena encompass all agents of physical change in the universe. Moreover, they are considered to be fundamental building blocks which developed the universe, and which is responsible for the success of all life on earth. However, the scope here is limited to the relationship of transport phenomena to artificial engineered systems.

Overview

In physics, transport phenomena are all irreversible processes of statistical nature stemming from the random continuous motion of molecules, mostly observed in fluids. Every aspect of transport phenomena is grounded in two primary concepts : the conservation laws, and the constitutive equations. The conservation laws, which in the context of transport phenomena are formulated as continuity equations, describe how the quantity being studied must be conserved. The constitutive equations describe how the quantity in question responds to various stimuli via transport. Prominent examples include Fourier's Law of Heat Conduction and the Navier-Stokes equations, which describe, respectively, the response of heat flux to temperature gradients and the relationship between fluid flux and the forces applied to the fluid. These equations also demonstrate the deep connection between transport phenomena and thermodynamics, a connection that explains why transport phenomena are irreversible. Almost all of these physical phenomena ultimately involve systems seeking their lowest energy state in keeping with the principle of minimum energy. As they approach this state, they tend to achieve true thermodynamic equilibrium, at which point there are no longer any driving forces in the system and transport ceases. The various aspects of such equilibrium are directly connected to a specific transport: heat transfer is the system's attempt to achieve thermal equilibrium with its environment, just as mass and momentum transport move the system towards chemical and mechanical equilibrium.

Examples of transport processes include heat conduction (energy transfer), fluid flow (momentum transfer), molecular diffusion (mass transfer), radiation and electric charge transfer in semiconductors.

Transport phenomena have wide application. For example, in solid state physics, the motion and interaction of electrons, holes and phonons are studied under "transport phenomena". Another example is in biomedical engineering, where some transport phenomena of interest are thermoregulation, perfusion, and microfluidics. In chemical engineering, transport phenomena are studied in reactor design, analysis of molecular or diffusive transport mechanisms, and metallurgy.

The transport of mass, energy, and momentum can be affected by the presence of external sources:

- An odor dissipates more slowly (and may intensify) when the source of the odor remains present.
- The rate of cooling of a solid that is conducting heat depends on whether a heat source is applied.

- The gravitational force acting on a rain drop counteracts the resistance or drag imparted by the surrounding air.

Commonalities Among Phenomena

An important principle in the study of transport phenomena is analogy between phenomena.

Diffusion

There are some notable similarities in equations for momentum, energy, and mass transfer which can all be transported by diffusion, as illustrated by the following examples:

- Mass: the spreading and dissipation of odors in air is an example of mass diffusion.
- Energy: the conduction of heat in a solid material is an example of heat diffusion.
- Momentum: the drag experienced by a rain drop as it falls in the atmosphere is an example of momentum diffusion (the rain drop loses momentum to the surrounding air through viscous stresses and decelerates).

The molecular transfer equations of Newton's law for fluid momentum, Fourier's law for heat, and Fick's law for mass are very similar. One can convert from one transfer coefficient to another in order to compare all three different transport phenomena.

A great deal of effort has been devoted in the literature to developing analogies among these three transport processes for turbulent transfer so as to allow prediction of one from any of the others. The Reynolds analogy assumes that the turbulent diffusivities are all equal and that the molecular diffusivities of momentum (μ/ρ) and mass (D_{AB}) are negligible compared to the turbulent diffusivities. When liquids are present and/or drag is present, the analogy is not valid. Other analogies, such as von Karman's and Prandtl's, usually result in poor relations.

The most successful and most widely used analogy is the Chilton and Colburn J-factor analogy. This analogy is based on experimental data for gases and liquids in both the laminar and turbulent regimes. Although it is based on experimental data, it can be shown to satisfy the exact solution derived from laminar flow over a flat plate. All of this information is used to predict transfer of mass.

Onsager Reciprocal Relations

In fluid systems described in terms of temperature, matter density, and pressure, it is known that temperature differences lead to heat flows from the warmer to the colder parts of the system; similarly, pressure differences will lead to matter flow from high-pressure to low-pressure regions (a "reciprocal relation"). What is remarkable is the observation that, when both pressure and temperature vary, temperature differences at constant pressure can cause matter flow (as in convection) and pressure differences at constant temperature can cause heat flow. Perhaps surprisingly, the heat flow per unit of pressure difference and the density (matter) flow per unit of temperature difference are equal.

This equality was shown to be necessary by Lars Onsager using statistical mechanics as a consequence of the time reversibility of microscopic dynamics. The theory developed by Onsager is much more general than this example and capable of treating more than two thermodynamic forces at once.

Momentum Transfer

In momentum transfer, the fluid is treated as a continuous distribution of matter. The study of momentum transfer, or fluid mechanics can be divided into two branches: fluid statics (fluids at rest), and fluid dynamics (fluids in motion). When a fluid is flowing in the x direction parallel to a solid surface, the fluid has x-directed momentum, and its concentration is $\upsilon_x\rho$. By random diffusion of molecules there is an exchange of molecules in the z direction. Hence the x-directed momentum has been transferred in the z-direction from the faster- to the slower-moving layer. The equation for momentum transport is Newton's Law of Viscosity written as follows:

$$\tau_{zx} = -\nu \frac{\partial \rho \upsilon_x}{\partial z}$$

where τ_{zx} is the flux of x-directed momentum in the z direction, ν is μ/ρ, the momentum diffusivity, z is the distance of transport or diffusion, ρ is the density, and μ is the viscosity. Newtons Law is the simplest relationship between the flux of momentum and the velocity gradient.

Mass Transfer

When a system contains two or more components whose concentration vary from point to point, there is a natural tendency for mass to be transferred, minimizing any concentration difference within the system. Mass Transfer in a system is governed by Fick's First Law: 'Diffusion flux from higher concentration to lower concentration is proportional to the gradient of the concentration of the substance and the diffusivity of the substance in the medium.' Mass transfer can take place due to different driving forces. Some of them are:

- Mass can be transferred by the action of a pressure gradient(pressure diffusion).
- Forced diffusion occurs because of the action of some external force.

- Diffusion can be caused by temperature gradients (thermal diffusion).
- Diffusion can be caused by differences in chemical potential.

This can be compared to Fick's Law of Diffusion:

$$J_{Ay} = -D_{AB}\frac{\partial Ca}{\partial y}$$

where D is the diffusivity constant.

Energy Transfer

All processes in engineering involve the transfer of energy. Some examples are the heating and cooling of process streams, phase changes, distillations, etc. The basic principle is the first law of thermodynamics which is expressed as follows for a static system:

$$q = -k\frac{dT}{dx}$$

The net flux of energy through a system equals the conductivity times the rate of change of temperature with respect to position.

For other systems that involve either turbulent flow, complex geometries or difficult boundary conditions another equation would be easier to use:

$$Q = h \cdot A \cdot \Delta T$$

where A is the surface area, : ΔT is the temperature driving force, Q is the heat flow per unit time, and h is the heat transfer coefficient.

Within heat transfer, two types of convection can occur:

Forced convection can occur in both laminar and turbulent flow. In the situation of laminar flow in circular tubes, several dimensionless numbers are used such as Nusselt number, Reynolds number, and Prandtl. The commonly used equation is:

$$Nu_a = \frac{h_a D}{k}$$

Natural or free convection is a function of Grashof and Prandtl numbers. The complexities of free convection heat transfer make it necessary to mainly use empirical relations from experimental data.

Heat transfer is analyzed in packed beds, reactors and heat exchangers.

Stefan Tube

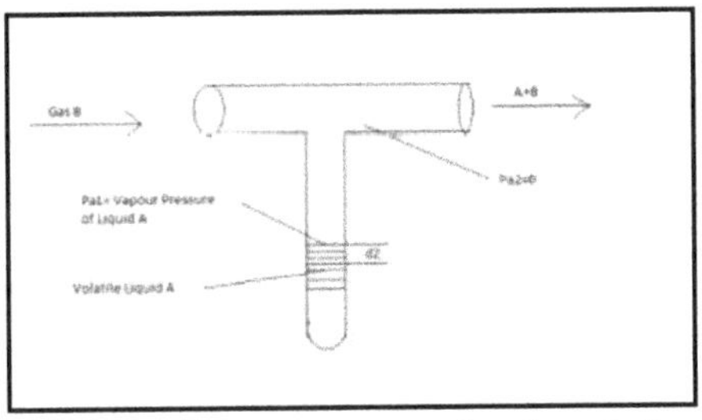

A schematic of a stefan tube. Note that it is not necessary for there to be a cross-wise horizontal tube at the top.

In chemical engineering, a Stefan tube is a device that was devised by Josef Stefan in 1874. It is often used for measuring diffusion coefficients. It comprises a vertical tube, over the top of which a gas flows and at the bottom of which is a pool of volatile liquid that is maintained in a constant-temperature bath. The liquid in the pool evaporates, diffuses through the gas above it in the tube, and is carried away by the gas flow over the tube mouth at the top. One then measures the fall in the level of the liquid in the tube.

The tube conventionally has a narrow diameter, in order to suppress convection.

The way that a Stefan tube is modelled, mathematically, is very similar to how one can model the diffusion of perfume fragrance molecules from (say) a drop of perfume on skin or clothes, evaporating up through the air to a person's nose. There are some differences between the models. However, they turn out to have little effect on results at highly dilute vapour concentrations.

Analysis

In the analysis of the system, various assumptions are made. The liquid, conventionally denoted A, is neither soluble in the gas in the tube, conventionally denoted B, nor reacts with it. The decrease in volume of the liquid A and increase in volume of the gas B over time can be ignored for the purposes of solving the equations that describe the behaviour, and an assumption can be made that the instantaneous flux at any time is the steady state value. There are no radial or circumferential components to the concentration gradients, resulting from convection or turbulence caused by excessively vigorous flow at the upper mouth of the tube, and the diffusion can thus be treated as a simple one-dimensional flow in the vertical direction. The mole fraction of A at the upper mouth of the tube is zero, as a consequence of the gas flow. At the interface between A and B the flux of B is zero (because it is insoluble in A) and the mole fraction is the equilibrium value.

The flux of B, denoted N_B, is thus zero throughout the tube, its diffusive flux downward (along its concentration gradient) is balanced by its convective flux upward caused by A.

Applying these assumptions, the system can be modelled using Fick's laws of diffusion or as Maxwell–Stefan diffusion.

Chemical Reactor

In chemical engineering, chemical reactors are vessels designed to contain chemical reactions. Also referred to as a reaction vessel, the reactants contained are substances that change form after a chemical reaction. One example is a pressure reactor. The design of a chemical reactor deals with multiple aspects of chemical engineering. Chemical engineers design reactors to maximize net present value for the given reaction. Designers ensure that the reaction proceeds with the highest efficiency towards the desired output product, producing the highest yield of product while requiring the least amount of money to purchase and operate. Normal operating expenses include energy input, energy removal, raw material costs, labor, etc. Energy changes can come in the form of heating or cooling, pumping to increase pressure, frictional pressure loss (such as pressure drop across a 90° elbow or an orifice plate) or agitation.

Chemical reaction engineering is the branch of chemical engineering which deals with chemical reactors and their design, especially by application of chemical kinetics to industrial systems.

Overview

Cut-away view of a stirred-tank chemical reactor with a cooling jacket.

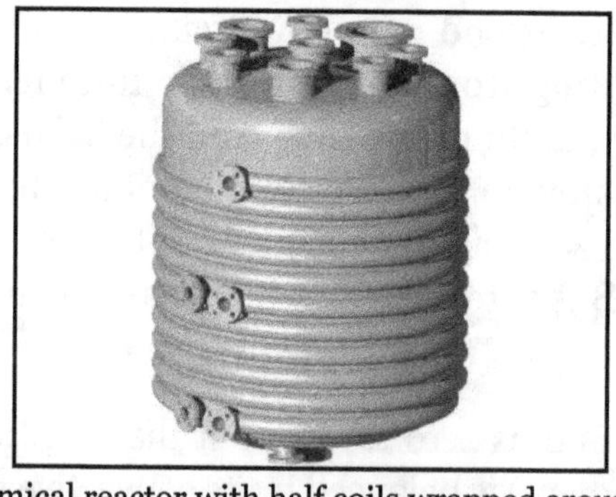

Chemical reactor with half coils wrapped around it.

There are a couple of main basic vessel types:

- A tank,
- A pipe or tubular reactor (laminar flow reactor(LFR)).

Both types can be used as continuous reactors or batch reactors, and either may accommodate one or more solids (reagents, catalyst, or inert materials), but the reagents and products are typically fluids. Most commonly, reactors are run at steady-state, but can also be operated in a transient state. When a reactor is first brought into operation (after maintenance or inoperation) it would be considered to be in a transient state, where key process variables change with time.

There are three main basic models used to estimate the most important process variables of different chemical reactors:

- *batch reactor* model (batch),
- *continuous stirred-tank reactor* model (CSTR),
- *plug flow reactor* model (PFR).

Furthermore, catalytic reactors require separate treatment, whether they are batch, CST, or PF reactors, as the many assumptions of the simpler models are not valid.

Key process variables include:

- Residence time (τ, lower case Greek tau),
- Volume (V),
- Temperature (T),
- Pressure (P),
- Concentrations of chemical species (C_1, C_2, C_3, ... C_n),
- Heat transfer coefficients (h, U),
- Chemical reactor manufacturer (h, U).

A chemical reactor, typically tubular reactor, could be a packed bed. Typical packed bed reactors consist of a chamber, such as a tube or channel that contains catalyst particles or pellets, and a liquid that flows through the catalyst. A chemical reactor may also be a fluidized bed.

Chemical reactions occurring in a reactor may be exothermic, meaning giving off heat, or endothermic, meaning absorbing heat. A chemical reactor vessel may have a cooling or heating jacket or cooling or heating coils (tubes) wrapped around the outside of its vessel wall to cool down or heat up the contents.

Types

Batch Reactor

The simplest type of reactor is a batch reactor. Materials are charged in a batch reactor, and the reaction proceeds with time. A batch reactor does not reach a steady state, and control of temperature, pressure and volume is necessary. Thus, a batch reactor has ports for sensors and material input and output. Batch reactors are used in small-scale production and reactions with biological materials, such as in brewing, pulping and production of enzymes.

CSTR (Continuous Stirred-tank Reactor)

Checking condition inside the case of a continuous stirred tank reactor (CSTR). Note the impeller (or agitator) blades on the shaft for mixing. Also note the baffle at the bottom of the image which also helps in mixing.

In a CSTR, one or more fluid reagents are introduced into a tank reactor (typically) equipped with an impeller while the reactor effluent is removed. The impeller stirs the reagents to ensure proper mixing. Simply dividing the volume of the tank by the average volumetric flow rate through the tank gives the *space time*, or the average amount of time a discrete quantity of reagent spends inside the tank. Using chemical kinetics, the reaction's expected percent completion can be calculated. Some important aspects of the CSTR:

- At steady-state, the mass flow rate in must equal the mass flow rate out, otherwise the tank will overflow or go empty (transient state). While the reactor is in a transient state the model equation must be derived from the differential mass and energy balances.
- The reaction proceeds at the reaction rate associated with the final (output) concentration, since the concentration is assumed to be homogenous throughout the reactor.

- Often, it is economically beneficial to operate several CSTRs in series. This allows, for example, the first CSTR to operate at a higher reagent concentration and therefore a higher reaction rate. In these cases, the sizes of the reactors may be varied in order to minimize the total capital investment required to implement the process.
- It can be demonstrated that an infinite number of infinitely small CSTRs operating in series would be equivalent to a PFR.

The behavior of a CSTR is often approximated or modeled by that of a Continuous Ideally Stirred-Tank Reactor (CISTR). All calculations performed with CISTRs assume perfect mixing. If the residence time is 5-10 times the mixing time, this approximation is considered valid for engineering purposes. The CISTR model is often used to simplify engineering calculations and can be used to describe research reactors. In practice it can only be approached, particularly in industrial size reactors in which the mixing time may be very large.

A loop reactor is a hybrid type of catalytic reactor that physically resembles a tubular reactor, but operates like a CSTR. The reaction mixture is circulated in a loop of tube, surrounded by a jacket for cooling or heating, and there is a continuous flow of starting material in and product out.

PFR (Plug Flow Reactor)

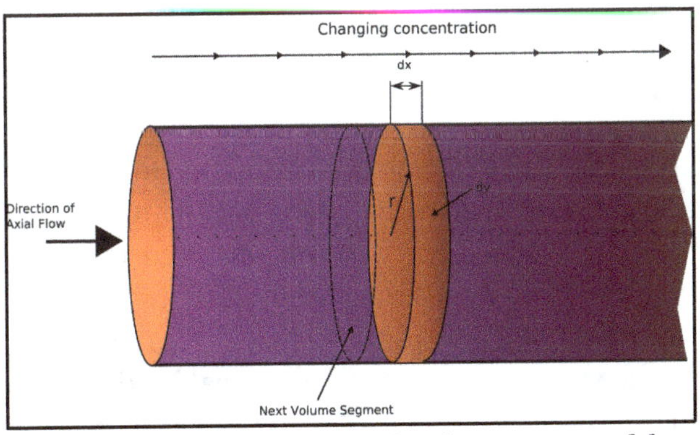

Simple diagram illustrating plug flow reactor model.

In a PFR, sometimes called continuous tubular reactor (CTR), one or more fluid reagents are pumped through a pipe or tube. The chemical reaction proceeds as the reagents travel through the PFR. In this type of reactor, the changing reaction rate creates a gradient with respect to distance traversed; at the inlet to the PFR the rate is very high, but as the concentrations of the reagents decrease and the concentration of the product(s) increases the reaction rate slows. Some important aspects of the PFR:

- All calculations performed with PFRs assume no upstream or downstream mixing, as implied by the term "plug flow".

- Reagents may be introduced into the PFR at locations in the reactor other than the inlet. In this way, a higher efficiency may be obtained, or the size and cost of the PFR may be reduced.
- A PFR typically has a higher efficiency than a CSTR of the same volume. That is, given the same space-time (or residence time), a reaction will typically proceed to a higher percentage completion in a PFR than in a CSTR. This is not always true for reversible reactions.

For most chemical reactions of industrial interest, it is impossible for the reaction to proceed to 100% completion. The rate of reaction decreases as the reactants are consumed until the point where the system reaches dynamic equilibrium (no net reaction, or change in chemical species occurs). The equilibrium point for most systems is less than 100% complete. For this reason a separation process, such as distillation, often follows a chemical reactor in order to separate any remaining reagents or byproducts from the desired product. These reagents may sometimes be reused at the beginning of the process, such as in the Haber process. In some cases, very large reactors would be necessary to approach equilibrium, and chemical engineers may choose to separate the partially reacted mixture and recycle the leftover reactants.

Continuous oscillatory baffled reactor (COBR) is a tubular plug flow reactor. The mixing in COBR is achieved by the combination of fluid oscillation and orifice baffles, allowing plug flow to be achieved under laminar flow conditions with the net flow Reynolds number just about 100.

Semi-batch Reactor

A semi-batch reactor is operated with both continuous and batch inputs and outputs. A fermenter, for example, is loaded with a batch of medium and microbes which constantly produces carbon dioxide that must be removed continuously. Analogously, driving a reaction of gas with a liquid is usually difficult, since the gas bubbles off. Therefore, a continuous feed of gas is injected into the batch of a liquid. One chemical reactant is charged to the vessel and a second chemical is added slowly (for instance, to prevent side reactions).

Catalytic Reactor

Although catalytic reactors are often implemented as plug flow reactors, their analysis requires more complicated treatment. The rate of a catalytic reaction is proportional to the amount of catalyst the reagents contact, as well as the concentration of the reactants. With a solid phase catalyst and fluid phase reagents, this is proportional to the exposed area, efficiency of diffusion of reagents in and products out, and efficacy of mixing. Perfect mixing usually cannot be assumed. Furthermore, a catalytic reaction pathway often occurs in multiple steps with intermediates that are chemically bound to the catalyst; and as the chemical binding to the catalyst is also a chemical reaction, it may affect the kinetics. Catalytic reactions often display the so-called *falsified kinetics*, i.e. the apparent kinetics differ from elementary chemical kinetics due to physical transport effects.

The behavior of the catalyst is also a consideration. Particularly in high-temperature petrochemical processes, catalysts are deactivated by sintering, coking, and similar processes.

A common example of a catalytic reactor is the catalytic converter following an engine. However, most petrochemical reactors are catalytic, and are responsible for most of industrial chemical production in the world, with extremely high-volume examples such as sulfuric acid, ammonia, reformate/BTEX (benzene, toluene, ethylbenzene and xylene) and alkylate gasoline blending stock.

Chemical Plant

BASF Chemical Plant Portsmouth Site in the West Norfolk area of Portsmouth, Virginia, United States. The plant is served by the Commonwealth Railway.

A chemical plant is an industrial process plant that manufactures (or otherwise processes) chemicals, usually on a large scale. The general objective of a chemical plant is to create new material wealth via the chemical or biological transformation and or separation of materials. Chemical plants use specialized equipment, units, and technology in the manufacturing process. Other kinds of plants, such as polymer, pharmaceutical, food, and some beverage production facilities, power plants, oil refineries or other refineries, natural gas processing and biochemical plants, water and wastewater treatment, and pollution control equipment use many technologies that have similarities to chemical plant technology such as fluid systems and chemical reactor systems. Some would consider an oil refinery or a pharmaceutical or polymer manufacturer to be effectively a chemical plant.

Petrochemical plants (plants using chemicals from petroleum as a raw material or *feedstock*) are usually located adjacent to an oil refinery to minimize transportation costs for the feedstocks produced by the refinery. Speciality chemical and fine chemical plants are usually much smaller and not as sensitive to location. Tools have been developed for converting a base project cost from one geographic location to another.

Chemical Processes

Chemical plants use chemical processes, which are detailed industrial-scale methods, to transform feedstock chemicals into products. The same chemical process can be used at more than one chemical plant, with possibly differently scaled capacities at each plant. Also, a chemical plant at a site may be constructed to utilize more than one chemical process, for instance to produce multiple products.

A chemical plant commonly has usually large vessels or sections called units or lines that are interconnected by piping or other material-moving equipment which can carry streams of material. Such material streams can include fluids (gas or liquid carried in piping) or sometimes solids or mixtures such as slurries. An overall chemical process is commonly made up of steps called unit operations which occur in the individual units. A raw material going into a chemical process or plant as input to be converted into a product is commonly called a feedstock, or simply feed. In addition to feedstocks for the plant as a whole, an input stream of material to be processed in a particular unit can similarly be considered feed for that unit. Output streams from the plant as a whole are final products and sometimes output streams from individual units may be considered intermediate products for their units. However, final products from one plant may be intermediate chemicals used as feedstock in another plant for further processing. For example, some products from an oil refinery may used as feedstock in petrochemical plants, which may in turn produce feedstocks for pharmaceutical plants.

Either the feedstock(s), the product(s), or both may be individual compounds or mixtures. It is often not worthwhile separating the components in these mixtures completely; specific levels of purity depend on product requirements and process economics.

Operations

Chemical processes may be run in continuous or batch operation.

Batch Operation

In batch operation, production occurs in time-sequential steps in discrete batches. A batch of feedstock(s) is fed (or charged) into a process or unit, then the chemical process takes place, then the product(s) and any other outputs are removed. Such batch production may be repeated over again and again with new batches of feedstock. Batch operation is commonly used in smaller scale plants such as pharmaceutical or specialty chemicals production, for purposes of improved traceability as well as flexibility. Continuous plants are usually used to manufacture commodity or petrochemicals while batch plants are more common in speciality and fine chemical production as well as pharmaceutical active ingredient (API) manufacture.

Continuous Operation

In continuous operation, all steps are ongoing continuously in time. During usual continuous operation, the feeding and product removal are ongoing streams of moving material, which together with the process itself, all take place simultaneously and continuously. Chemical plants or units in continuous operation are usually in a steady state or approximate steady state. Steady state means that quantities related to the process do not change as time passes during operation. Such constant quantities include stream flow rates, heating or cooling rates, temperatures, pressures, and chemical compositions at any given point (location). Continuous operation is more efficient in many large scale operations like petroleum refineries. It is possible for some units to operate continuously and others be in batch operation in a chemical plant. The amount of primary feedstock or product per unit of time which a plant or unit can process is referred to as the capacity of that plant or unit. For examples: the capacity of an oil refinery may be given in terms of barrels of crude oil refined per day; alternatively chemical plant capacity may be given in tons of product produced per day. In actual daily operation, a plant (or unit) will operate at a percentage of its full capacity. Engineers typically assume 90% operating time for plants which work primarily with fluids, and 80% uptime for plants which primarily work with solids.

Units and Fluid Systems

Specific unit operations are conducted in specific kinds of units. Although some units may operate at ambient temperature or pressure, many units operate at higher or lower temperatures or pressures. Vessels in chemical plants are often cylindrical with rounded ends, a shape which can be suited to hold either high pressure or vacuum. Chemical reactions can convert certain kinds of compounds into other compounds in chemical reactors. Chemical reactors may be packed beds and may have solid heterogeneous catalysts which stay in the reactors as fluids move through, or may be simply be stirred vessels in which reactions occur. Since the surface of solid heterogeneous catalysts may sometimes become "poisoned" from deposits such as coke, regeneration of catalysts may be necessary. Fluidized beds may also be used in some cases to ensure good mixing. There can also be units (or subunits) for mixing (including dissolving), separation, heating, cooling, or some combination of these. For example, chemical reactors often have stirring for mixing and heating or cooling to maintain temperature. When designing plants on a large scale, heat produced or absorbed by chemical reactions must be considered. Some plants may have units with organism cultures for biochemical processes such as fermentation or enzyme production.

Distillation unit in Italy.

Separation processes include filtration, settling (sedimentation), extraction or leaching, distillation, recrystallization or precipitation (followed by filtration or settling), reverse osmosis, drying, and adsorption. Heat exchangers are often used for heating or cooling, including boiling or condensation, often in conjunction with other units such as distillation towers. There may also be storage tanks for storing feedstock, intermediate or final products, or waste. Storage tanks commonly have level indicators to show how full they are. There may be structures holding or supporting sometimes massive units and their associated equipment. There are often stairs, ladders, or other steps for personnel to reach points in the units for sampling, inspection, or maintenance. An area of a plant or facility with numerous storage tanks is sometimes called a *tank farm*, especially at an oil depot.

Fluid systems for carrying liquids and gases include piping and tubing of various diameter sizes, various types of valves for controlling or stopping flow, pumps for moving or pressurizing liquid, and compressors for pressurizing or moving gases. Vessels, piping, tubing, and sometimes other equipment at high or very low temperature are commonly covered with insulation for personnel safety and to maintain temperature inside. Fluid systems and units commonly have instrumentation such as temperature and pressure sensors and flow measuring devices at select locations in a plant. Online analyzers for chemical or physical property analysis have become more common. Solvents can sometimes be used to dissolve reactants or materials such as solids for extraction or leaching, to provide a suitable medium for certain chemical reactions to run, or so they can otherwise be treated as fluids.

Chemical Plant Design

Today, the fundamental aspects of designing chemical plants are done by chemical engineers. Historically, this was not always the case and many chemical plants were constructed in a haphazard way before the discipline of chemical engineering became established. Chemical engineering was first established as a profession in the United Kingdom when the first chemical engineering course was given at the University of Manchester in 1887 by George E. Davis in the form of twelve lectures covering various aspects of industrial chemical practice. As a consequence George E. Davis is regarded as the World's first Chemical Engineer.Today Chemical Engineering is a profession and those Professional Chemical Engineers with experience can gain "Chartered" engineer status through the Institution of Chemical Engineers.

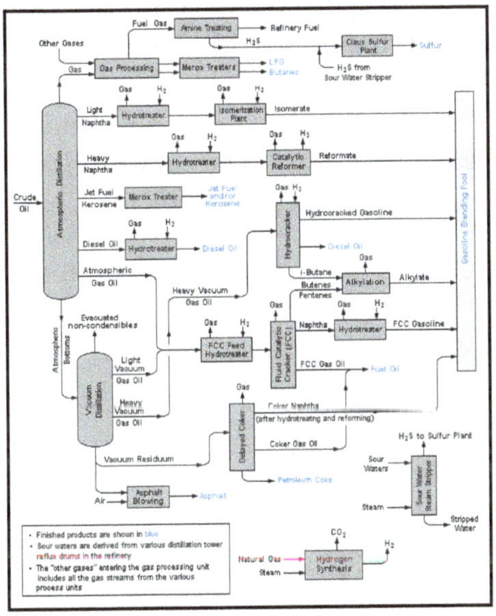

Flow diagram for a typical oil refinery.

In plant design, typically less than 1 per cent of ideas for new designs ever become commercialized. During this solution process, typically, cost studies are used as an initial screening to eliminate unprofitable designs. If a process appears profitable, then other factors are considered, such as safety, environmental constraints, controllability, etc. The general goal in plant design, is to construct or synthesize "optimum designs" in the neighborhood of the desired constraints.

Many times chemists research chemical reactions or other chemical principles in a laboratory, commonly on a small scale in a "batch-type" experiment. Chemistry information obtained is then used by chemical engineers, along with expertise of their own, to convert to a chemical process and scale up the batch size or capacity. Commonly, a small chemical plant called a pilot plant is built to provide design and operating information before construction of a large plant. From data and operating experience obtained from the pilot plant, a scaled-up plant can be designed for higher or full capacity. After the fundamental aspects of a plant design are determined, mechanical or electrical engineers may become involved with mechanical or electrical details, respectively. Structural engineers may become involved in the plant design to ensure the structures can support the weight of the units, piping, and other equipment.

The units, streams, and fluid systems of chemical plants or processes can be represented by block flow diagrams which are very simplified diagrams, or process flow diagrams which are somewhat more detailed. The streams and other piping are shown as lines with arrow heads showing usual direction of material flow. In block diagrams, units are often simply shown as blocks. Process flow diagrams may use more detailed symbols and show pumps, compressors, and major valves. Likely values or ranges of material flow rates for the various streams are determined based on desired plant capacity using material balance calculations. Energy balances are also done based on heats of reaction, heat capacities, expected temperatures and pressures at various points to calculate amounts of heating and cooling needed in various places and to size heat exchangers. Chemical plant design can be shown in fuller detail in a piping and instrumentation diagram (P&ID) which shows all piping, tubing, valves, and instrumentation, typically with special symbols. Showing a full plant is often complicated in a P&ID, so often only individual units or specific fluid systems are shown in a single P&ID.

In the plant design, the units are sized for the maximum capacity each may have to handle. Similarly, sizes for pipes, pumps, compressors, and associated equipment are chosen for the flow capacity they have to handle. Utility systems such as electric power and water supply should also be included in the plant design. Additional piping lines for non-routine or alternate operating procedures, such as plant or unit startups and shutdowns, may have to be included. Fluid systems design commonly includes isolation valves around various units or parts of a plant so that a section of a plant could be isolated in case of a problem such as a leak in a unit. If pneumatically or hydraulically actuated valves are used, a system of pressurizing lines to the actuators is needed. Any points where process samples may have to be taken should have sampling lines, valves, and access to them included in the detailed design. If necessary, provisions should be made for reducing high pressure or temperature of a sampling stream, such including a pressure reducing valve or sample cooler.

Units and fluid systems in the plant including all vessels, piping, tubing, valves, pumps, compressors, and other equipment must be rated or designed to be able to withstand the entire range of pressures, temperatures, and other conditions which they could possibly encounter, including any appropriate safety factors. All such units and equipment should also be checked for materials compatibility to ensure they can withstand long-term exposure to the chemicals they will come in contact with. Any closed system in a plant which has a means of pressurizing possibly beyond the rating of its equipment, such as heating, exothermic reactions, or certain pumps or compressors, should have an appropriately sized pressure relief valve included to prevent overpressurization for safety. Frequently all of these parameters (temperatures, pressures, flow, etc.) are exhaustively analyzed in combination through a *Hazop* or *fault tree analysis*, to ensure that the plant has no known risk of serious hazard.

Within any constraints the plant is subject to, design parameters are optimized for good economic performance while ensuring safety and welfare of personnel and the surrounding community. For flexibility, a plant may be designed to operate in a range around some optimal design parameters in case feedstock or economic conditions change and re-optimization is desirable. In more modern times, computer simulations or other computer calculations have been used to help in chemical plant design or optimization.

Plant Operation

Process Control

In process control, information gathered automatically from various sensors or other devices in the plant is used to control various equipment for running the plant, thereby controlling operation of the plant. Instruments receiving such information signals and sending out control signals to perform this function automatically are process *controllers*. Previously, pneumatic controls were sometimes used. Electrical controls are now common. A plant often has a control room with displays of parameters such as key temperatures, pressures, fluid flow rates and levels, operating positions of key valves, pumps and other equipment, etc. In addition, operators in the control room can control various aspects of the plant operation, often including overriding automatic control. Process control with a computer represents more modern technology. Based on possible changing feedstock composition, changing products requirements or economics, or other changes in constraints, operating conditions may be re-optimized to maximize profit.

Workers

Workers in Italy, 1969. Photo by Paolo Monti.

As in any industrial setting, there are a variety of workers working throughout a chemical plant facility, often organized into departments, sections, or other work groups.

Such workers typically include engineers, plant operators, and maintenance technicians. Other personnel at the site could include chemists, management/administration and office workers. Types of engineers involved in operations or maintenance may include chemical process engineers, mechanical engineers for maintaining mechanical equipment, and electrical/computer engineers for electrical or computer equipment.

Transport

Large quantities of fluid feedstock or product may enter or leave a plant by pipeline, railroad tank car, or tanker truck. For example, petroleum commonly comes to a refinery by pipeline. Pipelines can also carry petrochemical feedstock from a refinery to a nearby petrochemical plant. Natural gas is a product which comes all the way from a natural gas processing plant to final consumers by pipeline or tubing. Large quantities of liquid feedstock are typically pumped into process units. Smaller quantities of feedstock or product may be shipped to or from a plant in drums. Use of drums about 55 gallons in capacity is common for packaging industrial quantities of chemicals. Smaller batches of feedstock may be added from drums or other containers to process units by workers.

Maintenance

In addition to feeding and operating the plant, and packaging or preparing the product for shipping, plant workers are needed for taking samples for routine and troubleshooting analysis and for performing routine and non-routine maintenance. Routine maintenance can include periodic inspections and replacement of worn catalyst, analyzer reagents, various sensors, or mechanical parts. Non-routine maintenance can include investigating problems and then fixing them, such as leaks, failure to meet feed or product specifications, mechanical failures of valves, pumps, compressors, sensors, etc.

Statutory and Regulatory Compliance

When working with chemicals, safety is a concern in order to avoid problems such as chemical accidents . In the United States, the law requires that employers provide workers working with chemicals with access to a Material Safety Data Sheet (MSDS) for every kind of chemical they work with. An MSDS for a certain chemical is prepared and provided by the supplier to whoever buys the chemical. Other laws covering chemical safety, hazardous waste, and pollution must be observed, including statutes such as the Resource Conservation and Recovery Act (RCRA) and the Toxic Substances Control Act (TSCA), and regulations such as the Chemical Facility Anti-Terrorism Standards in the United States. Hazmat (hazardous materials) teams are trained to deal with chemical leaks or spills. Process Hazard Analysis (PHA) is used to assess potential hazards in chemical plants. In 1998, the U. S. Chemical Safety and Hazard Investigation Board has become operational.

Plant Facilities

The actual production or process part of a plant may be indoors, outdoors, or a combination of the two. It may be a traditional stick-built plant or a modular skid. Large modular skids are especially impressive feats of engineering. A modular skid is built including all of the modular equipment needed to do the same job a traditional stick-build plant may perform. However, the modular skid is built within a structural steel frame, allowing it to be shipped to the onsite location without needing to be rebuilt onsite. A modular skid build results in a higher functioning end product, as less hands are required in the onsite setup of the modular skid process unit, resulting in minimized risk for mishaps. The actual production section of a facility usually has the appearance of a rather industrial environment. Hard hats and work shoes are commonly worn. Floors and stairs are often made of metal grating, and there is practically no decoration. There may also be pollution control or waste treatment facilities or equipment. Sometimes existing plants may be expanded or modified based on changing economics, feedstock, or product needs. As in other production facilities, there may be shipping and receiving, and storage facilities. In addition, there are usually certain other facilities, typically indoors, to support production at the site.

Although some simple sample analysis may be able to be done by operations technicians in the plant area, a chemical plant typically has a laboratory where chemists analyze samples taken from the plant. Such analysis can include chemical analysis or determination of physical properties. Sample analysis can include routine quality control on feedstock coming into the plant, intermediate and final products to ensure quality specifications are met. Non-routine samples may be taken and analyzed for investigating plant process problems also. A larger chemical company often has a research laboratory for developing and testing products and processes where there may be pilot plants, but such a laboratory may be located at a site separate from the production plants.

A plant may also have a workshop or maintenance facility for repairs or keeping maintenance equipment. There is also typically some office space for engineers, management or administration, and perhaps for receiving visitors. The decorum there is commonly more typical of an office environment.

Clustering of Commodity Chemical Plants

Chemical Plants used particularly for commodity chemical and petrochemical manufacture,are located in relatively few manufacturing locations around the world largely due to infrastructural needs.This is less important for speciality or fine chemical batch plants. Not all commodity/petrochemicals are produced in any one location but groups of related materials often are, to induce industrial symbiosis as well as material, energy and utility efficiency and other economies of scale. These manufacturing locations often have business clusters of units called chemical plants that share utilities and large scale infrastructure such as power stations, port facilities, road and rail terminals. In the United Kingdom for example there are four main locations for commodity chemical manufacture: near the River Mersey in Northwest England, on the Humber on the East coast of Yorkshire, in Grangemouth near the Firth of Forth in Scotland and on Teesside as part of the Northeast of England Process Industry Cluster (NEPIC). Approximately 50% of the UK's petrochemicals, which are also commodity chemicals, are produced by the industry cluster companies on Teesside at the mouth of the River Tees on three large chemical parks at Wilton, Billingham and Seal Sands.

Corrosion and use of New Materials

Corrosion in chemical process plants is a major issue that consumes billions of dollars yearly. Electrochemical corrosion of metals is pronounced in chemical process plants due to the presence of acid fumes and other electrolytic interactions. Recently, FRP (Fibre-reinforced plastic) is used as a material of construction. The British standard specification BS4994 is widely used for design and construction of the vessels, tanks, etc.

Chemical Thermodynamics

Chemical thermodynamics is the study of the interrelation of heat and work with chemical reactions or with physical changes of state within the confines of the laws of thermodynamics. Chemical thermodynamics involves not only laboratory measurements of various thermodynamic properties, but also the application of mathematical methods to the study of chemical questions and the *spontaneity* of processes.

The structure of chemical thermodynamics is based on the first two laws of thermodynamics. Starting from the first and second laws of thermodynamics, four equations called the "fundamental equations of Gibbs" can be derived. From these four, a multitude of equations, relating the thermodynamic properties of the thermodynamic system can be derived using relatively simple mathematics. This outlines the mathematical framework of chemical thermodynamics.

History

In 1865, the German physicist Rudolf Clausius, in his *Mechanical Theory of Heat*, suggested that the principles of thermochemistry, e.g. the heat evolved in combustion reactions, could be applied to the principles of thermodynamics. Building on the work of Clausius, between the years 1873-76 the American mathematical physicist Willard Gibbs published a series of three papers, the most famous one being the paper *On the Equilibrium of Heterogeneous Substances*. In these papers, Gibbs showed how the first two laws of thermodynamics could be measured graphically and mathematically to determine both the thermodynamic equilibrium of chemical reactions as well as their tendencies to occur or proceed. Gibbs' collection of papers provided the first unified body of thermodynamic theorems from the principles developed by others, such as Clausius and Sadi Carnot.

J. Willard Gibbs - founder of *chemical thermodynamics*.

During the early 20th century, two major publications successfully applied the principles developed by Gibbs to chemical processes, and thus established the foundation of the science of chemical thermodynamics. The first was the 1923 textbook *Thermodynamics and the Free Energy of Chemical Substances* by Gilbert N. Lewis and Merle Randall. This book was responsible for supplanting the chemical affinity with the term free energy in the English-speaking world. The second was the 1933 book *Modern Thermodynamics by the methods of Willard Gibbs* written by E. A. Guggenheim. In this manner, Lewis, Randall, and Guggenheim are considered as the founders of modern chemical thermodynamics because of the major contribution of these two books in unifying the application of thermodynamics to chemistry.

Overview

The primary objective of chemical thermodynamics is the establishment of a criterion for the determination of the feasibility or spontaneity of a given transformation. In this manner, chemical thermodynamics is typically used to predict the energy exchanges that occur in the following processes:

- Chemical reactions.
- Phase changes.
- The formation of solutions.

The following state functions are of primary concern in chemical thermodynamics:

- Internal energy (U).

- Enthalpy (*H*).
- Entropy (*S*).
- Gibbs free energy (*G*).

Most identities in chemical thermodynamics arise from application of the first and second laws of thermodynamics, particularly the law of conservation of energy, to these state functions.

The 3 Laws of Thermodynamics:

- The energy of the universe is constant.
- In any spontaneous process, there is always an increase in entropy of the universe.
- The entropy of a perfect crystal(well ordered) at 0 Kelvin is zero.

Chemical Energy

Chemical energy is the potential of a chemical substance to undergo a transformation through a chemical reaction or to transform other chemical substances. Breaking or making of chemical bonds involves energy or heat, which may be either absorbed or evolved from a chemical system.

Energy that can be released (or absorbed) because of a reaction between a set of chemical substances is equal to the difference between the energy content of the products and the reactants. This change in energy is called the change in internal energy of a chemical reaction. Where $\Delta U^{\circ}_{f\,\mathrm{reactants}}$ is the internal energy of formation of the reactant molecules that can be calculated from the bond energies of the various chemical bonds of the molecules under consideration and $\Delta U^{\circ}_{f\,\mathrm{products}}$ is the internal energy of formation of the product molecules. The change in internal energy is a process which is equal to the heat change if it is measured under conditions of constant volume (at STP condition), as in a closed rigid container such as a bomb calorimeter. However, under conditions of constant pressure, as in reactions in vessels open to the atmosphere, the measured heat change is not always equal to the internal energy change, because pressure-volume work also releases or absorbs energy. (The heat change at constant pressure is called the enthalpy change; in this case the enthalpy of formation).

Another useful term is the heat of combustion, which is the energy released due to a combustion reaction and often applied in the study of fuels. Food is similar to hydrocarbon fuel and carbohydrate fuels, and when it is oxidized, its caloric content is similar In chemical thermodynamics the term used for the chemical potential energy is chemical potential, and for chemical transformation an equation most often used is the Gibbs-Duhem equation.

Chemical Reactions

In most cases of interest in chemical thermodynamics there are internal degrees of freedom and processes, such as chemical reactions and phase transitions, which always create entropy unless they are at equilibrium, or are maintained at a "running equilibrium" through "quasi-static" changes by being coupled to constraining devices, such as pistons or electrodes, to deliver and receive external work. Even for homogeneous "bulk" materials, the free energy functions depend on the composition, as do all the extensive thermodynamic potentials, including the internal energy. If the quantities $\{N_i\}$, the number of chemical species, are omitted from the formulae, it is impossible to describe compositional changes.

Gibbs Function or Gibbs Energy

For a "bulk" (unstructured) system they are the last remaining extensive variables. For an unstructured, homogeneous "bulk" system, there are still various *extensive* compositional variables $\{N_i\}$ that G depends on, which specify the composition, the amounts of each chemical substance, expressed as the numbers of molecules present or (dividing by Avogadro's number = 6.023×10^{23}), the numbers of moles:

$$G = G(T, P, \{N_i\}).$$

For the case where only PV work is possible:

$$dG = -SdT + VdP + \sum_i \mu_i dN_i$$

in which μ_i is the chemical potential for the i-th component in the system:

$$\mu_i = \left(\frac{\partial G}{\partial N_i}\right)_{T,P,N_{j\neq i},etc.}.$$

The expression for dG is especially useful at constant T and P, conditions which are easy to achieve experimentally and which approximates the condition in living creatures:

$$(dG)_{T,P} = \sum_i \mu_i dN_i.$$

Chemical Affinity

While this formulation is mathematically defensible, it is not particularly transparent since one does not simply add or remove molecules from a system. There is always a *process* involved in changing the composition; e.g., a chemical reaction (or many), or movement of molecules from one phase (liquid) to another (gas or solid). We should find a notation which does not seem to imply that the amounts of the components (N_i) can be changed independently. All real processes obey conservation of mass, and in addition, conservation of the numbers of atoms of each kind. Whatever molecules are transferred to or from should be considered part of the "system".

Consequently, we introduce an explicit variable to represent the degree of advancement of a process, a progress variable ξ for the *extent of reaction* (Prigogine & Defay, p. 18; Prigogine, pp. 4–7; Guggenheim, p. 37.62), and to the use of the partial derivative $\partial G/\partial\xi$ (in place of the widely used "ΔG", since the quantity at issue is not a finite change). The result is an understandable expression for the dependence of dG on chemical reactions (or other processes). If there is just one reaction:

$$(dG)_{T,P} = \left(\frac{\partial G}{\partial \xi}\right)_{T,P} d\xi.$$

If we introduce the *stoichiometric coefficient* for the *i-th* component in the reaction:

$$\nu_i = \partial N_i / \partial \xi$$

which tells how many molecules of *i* are produced or consumed, we obtain an algebraic expression for the partial derivative:

$$\left(\frac{\partial G}{\partial \xi}\right)_{T,P} = \sum_i \mu_i \nu_i = -\mathbb{A}$$

where, (De Donder; Progoine & Defay, p. 69; Guggenheim, pp. 37,240), we introduce a concise and historical name for this quantity, the "affinity", symbolized by A, as introduced by Théophile de Donder in 1923. The minus sign comes from the fact the affinity was defined to represent the rule that spontaneous changes will ensue only when the change in the Gibbs free energy of the process is negative, meaning that the chemical species have a positive affinity for each other. The differential for G takes on a simple form which displays its dependence on compositional change:

$$(dG)_{T,P} = -\mathbb{A}d\xi.$$

If there are a number of chemical reactions going on simultaneously, as is usually the case:

$$(dG)_{T,P} = -\sum_k \mathbb{A}_k d\xi_k.$$

a set of reaction coordinates { ξ_j }, avoiding the notion that the amounts of the components (N_i) can be changed independently. The expressions above are equal to zero at thermodynamic equilibrium, while in the general case for real systems, they are negative because all chemical reactions proceeding at a finite rate produce entropy. This can be made even more explicit by introducing the reaction *rates* $d\xi_j/dt$. For each and every *physically independent process* (Prigogine & Defay, p. 38; Prigogine, p. 24),

$$\mathbb{A}\,\dot{\xi} \le 0.$$

This is a remarkable result since the chemical potentials are intensive system variables, depending only on the local molecular milieu. They cannot "know" whether the temperature and pressure (or any other system variables) are going to be held constant over time. It is a purely local criterion and must hold regardless of any such constraints. Of course, it could have been obtained by taking partial derivatives of any of the other fundamental state functions, but nonetheless is a general criterion for ($-T$ times) the entropy production from that spontaneous process; or at least any part of it that is not captured as external work.

We now relax the requirement of a homogeneous "bulk" system by letting the chemical potentials and the affinity apply to any locality in which a chemical reaction (or any other process) is occurring. By accounting for the entropy production due to irreversible processes, the inequality for dG is now replaced by an equality:

$$dG = -SdT + VdP - \sum_k \mathbb{A}_k d\xi_k + W'$$

or,

$$dG_{T,P} = -\sum_k \mathbb{A}_k d\xi_k + W'.$$

Any decrease in the Gibbs function of a system is the upper limit for any isothermal, isobaric work that can be captured in the surroundings, or it may simply be dissipated, appearing as T times a corresponding increase in the entropy of the system and/or its surrounding. Or it may go partly toward doing external work and partly toward creating entropy. The important point is that the *extent of reaction* for a chemical reaction may be coupled to the displacement of some external mechanical or electrical quantity in such a way that one can advance only if the other one also does. The coupling may occasionally be *rigid*, but it is often flexible and variable.

Solutions

In solution chemistry and biochemistry, the Gibbs free energy decrease ($\partial G/\partial\xi$, in molar units, denoted cryptically by ΔG) is commonly used as a surrogate for ($-T$ times) the entropy produced by spontaneous chemical reactions in situations where there is no work being done; or at least no "useful" work; i.e., other than perhaps some $\pm PdV$. The assertion that all *spontaneous reactions have a negative ΔG* is merely a restatement of the fundamental thermodynamic relation, giving it the physical dimensions of energy and somewhat obscuring its significance in terms of entropy. When there is no useful work being done, it would be less misleading to use the Legendre transforms of the entropy appropriate for constant T, or for constant T and P, the Massieu functions $-F/T$ and $-G/T$ respectively.

Non Equilibrium

Generally the systems treated with the conventional chemical thermodynamics are either at equilibrium or near equilibrium. Ilya Prigogine developed the thermodynamic treatment of open systems that are far from equilibrium. In doing so he has discovered phenomena and structures of completely new and completely unexpected types. His generalized, nonlinear and irreversible thermodynamics has found surprising applications in a wide variety of fields.

The non equilibrium thermodynamics has been applied for explaining how ordered structures e.g. the biological systems, can develop from disorder. Even if Onsager's relations are utilized, the classical principles of equilibrium in thermodynamics still show that linear systems close to equilibrium always develop into states of disorder which are stable to perturbations and cannot explain the occurrence of ordered structures.

Prigogine called these systems dissipative systems, because they are formed and maintained by the dissipative processes which take place because of the exchange of energy between the system and its environment and because they disappear if that exchange ceases. They may be said to live in symbiosis with their environment.

The method which Prigogine used to study the stability of the dissipative structures to perturbations is of very great general interest. It makes it possible to study the most varied problems, such as city traffic problems, the stability of insect communities, the development of ordered biological structures and the growth of cancer cells to mention but a few examples.

System Constraints

In this regard, it is crucial to understand the role of walls and other *constraints*, and the distinction between *independent* processes and *coupling*. Contrary to the clear implications of many reference sources, the previous analysis is not restricted to homogeneous, isotropic bulk systems which can deliver only PdV work to the outside world, but applies even to the most structured systems. There are complex systems with many chemical "reactions" going on at the same time, some of which are really only parts of the same, overall process. An *independent* process is one that *could* proceed even if all others were unaccountably stopped in their tracks. Understanding this is perhaps a "thought experiment" in chemical kinetics, but actual examples exist.

A gas reaction which results in an increase in the number of molecules will lead to an increase in volume at constant external pressure. If it occurs inside a cylinder closed with a piston, the equilibrated reaction can proceed only by doing work against an external force on the piston. The extent variable for the reaction can increase only if the piston moves, and conversely, if the piston is pushed inward, the reaction is driven backwards.

Similarly, a redox reaction might occur in an electrochemical cell with the passage of current in wires connecting the electrodes. The half-cell reactions at the electrodes are constrained if no current is allowed to flow. The current might be dissipated as joule heating, or it might in turn run an electrical device like a motor doing mechanical work. An automobile lead-acid battery can be recharged, driving the chemical reaction backwards. In this case as well, the reaction is not an independent process. Some, perhaps most, of the Gibbs free energy of reaction may be delivered as external work.

The hydrolysis of ATP to ADP and phosphate can drive the force times distance work delivered by living muscles, and synthesis of ATP is in turn driven by a redox chain in mitochondria and chloroplasts, which involves the transport of ions across the membranes of these cellular organelles. The coupling of processes here, and in the previous examples, is often not complete. Gas can leak slowly past a piston, just as it can slowly leak out of a rubber balloon. Some reaction may occur in a battery even if no external current is flowing. There is usually a coupling coefficient, which may depend on relative rates, which determines what percentage of the driving free energy is turned into external work, or captured as "chemical work"; a misnomer for the free energy of another chemical process.

Process Simulation

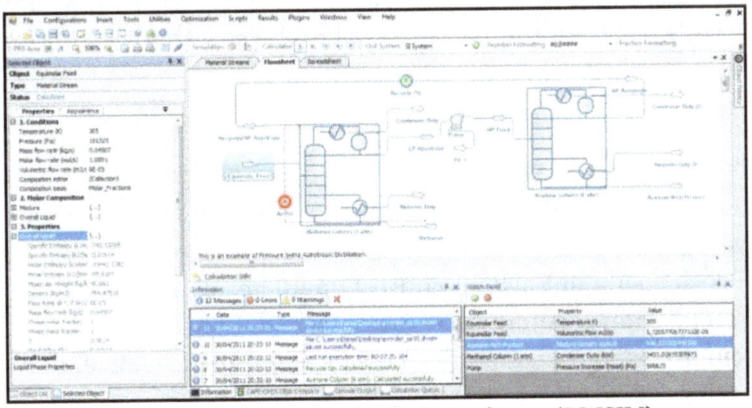

Screenshot of a process simulation software (DWSIM).

Process simulation is used for the design, development, analysis, and optimization of technical processes such as: chemical plants, chemical processes, environmental systems, power stations, complex manufacturing operations, biological processes, and similar technical functions.

Main Principle

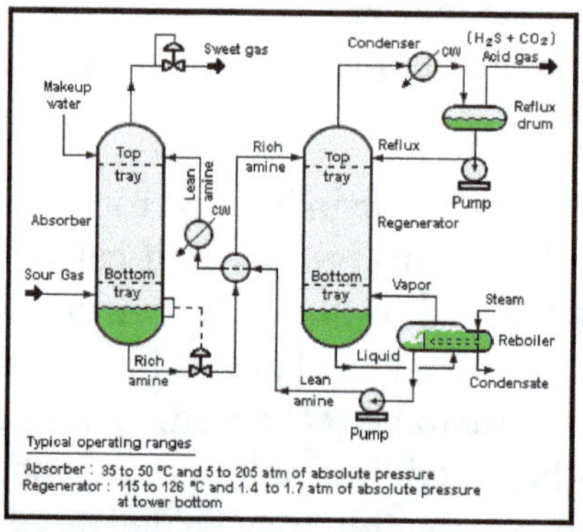

Process flow diagram of a typical amine treating process used in industrial plants.

Process simulation is a model-based representation of chemical, physical, biological, and other technical processes and unit operations in software. Basic prerequisites are a thorough knowledge of chemical and physical properties of pure components and mixtures, of reactions, and of mathematical models which, in combination, allow the calculation of a process in computers.

Process simulation software describes processes in flow diagrams where unit operations are positioned and connected by product or educt streams. The software has to solve the mass and energy balance to find a stable operating point. The goal of a process simulation is to find optimal conditions for an examined process. This is essentially an optimization problem which has to be solved in an iterative process.

Process simulation always use models which introduce approximations and assumptions but allow the description of a property over a wide range of temperatures and pressures which might not be covered by real data. Models also allow interpolation and extrapolation - within certain limits - and enable the search for conditions outside the range of known properties.

Modelling

The development of models for a better representation of real processes is the core of the further development of the simulation software. Model development is done on the chemical engineering side but also in control engineering and for the improvement of mathematical simulation techniques. Process simulation is therefore one of the few fields where scientists from chemistry, physics, computer science, mathematics, and several engineering fields work together.

VLE of the mixture of Chloroform and Methanol plus NRTL fit and extrapolation to different pressures.

A lot of efforts are made to develop new and improved models for the calculation of properties. This includes for example the description of:

- Thermophysical properties like vapor pressures, viscosities, caloric data, etc. of pure components and mixtures
- Properties of different apparatuses like reactors, distillation columns, pumps, etc.
- Chemical reactions and kinetics
- Environmental and safety-related data

Two main different types of models can be distinguished:

- Rather simple equations and correlations where parameters are fitted to experimental data.
- Predictive methods where properties are estimated.

The equations and correlations are normally preferred because they describe the property (almost) exactly. To obtain reliable parameters it is necessary to have experimental data which are usually obtained from factual data banks or, if no data are publicly available, from measurements.

Using predictive methods is much cheaper than experimental work and also than data from data banks. Despite this big advantage predicted properties are normally only used in early steps of the process development to find first approximate solutions and to exclude wrong pathways because these estimation methods normally introduce higher errors than correlations obtained from real data.

Process simulation also encouraged the further development of mathematical models in the fields of numerics and the solving of complex problems.

History

The history of process simulation is strongly related to the development of the computer science and of computer hardware and programming languages. Early working simple implementations of partial aspects of chemical processes were introduced in the 1970s when suitable hardware and software (here mainly the programming languages FORTRAN and C) became available. The modelling of chemical properties began much earlier, notably the cubic equation of states and the Antoine equation were precursory developments of the 19th century.

Steady State and Dynamic Process Simulation

Initially process simulation was used to simulate steady state processes. Steady-state models perform a mass and energy balance of a stationary process (a process in an equilibrium state) it does not depend on time.

Dynamic simulation is an extension of steady-state process simulation whereby time-dependence is built into the models via derivative terms i.e. accumulation of mass and energy. The advent of dynamic simulation means that the time-dependent description, prediction and control of real processes in real time has become possible. This includes the description of starting up and shutting down a plant, changes of conditions during a reaction, holdups, thermal changes and more.

Dynamic simulations require increased calculation time and are mathematically more complex than a steady state simulation. It can be seen as a multiply repeated steady state simulation (based on a fixed time step) with constantly changing parameters.

Dynamic simulation can be used in both an online and offline fashion. The online case being model predictive control, where the real-time simulation results are used to predict the changes that would occur for a control input change, and the control parameters are optimised based on the results. Offline process simulation can be used in the design, troubleshooting and optimisation of process plant as well as the conduction of case studies to assess the impacts of process modifications. Dynamic simulation is also used for operator training.

Chemical Kinetics

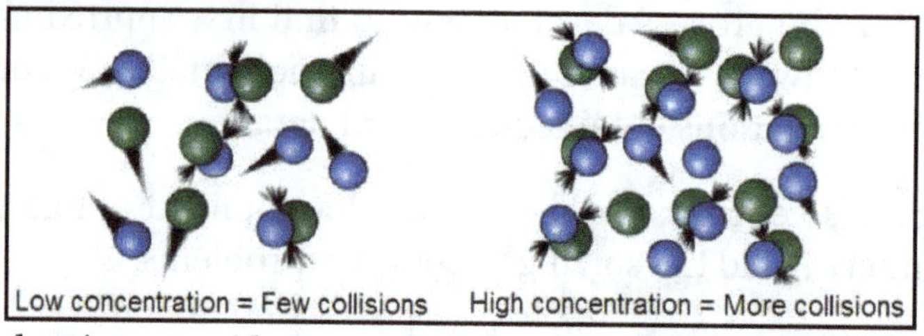

Reaction rate tends to increase with concentration – a phenomenon explained by collision theory.

Chemical kinetics, also known as reaction kinetics, is the study of rates of chemical processes. Chemical kinetics includes investigations of how different experimental conditions can influence the speed of a chemical reaction and yield information about the reaction's mechanism and transition states, as well as the construction of mathematical models that can describe the characteristics of a chemical reaction.

History

In 1864, Peter Waage and Cato Guldberg pioneered the development of chemical kinetics by formulating the law of mass action, which states that the speed of a chemical reaction is proportional to the quantity of the reacting substances.

Van 't Hoff studied chemical dynamics and published in 1884 his famous "Etudes de dynamique chimique". In 1901 he was awarded by the first Nobel Prize in Chemistry "in recognition of the extraordinary services he has rendered by the discovery of the laws of chemical dynamics and osmotic pressure in solutions". After van 't Hoff, chemical kinetics deals with the experimental determination of reaction rates from which rate laws and rate constants are derived. Relatively simple rate laws exist for zero order reactions (for which reaction rates are independent of concentration), first order reactions, and second order reactions, and can be derived for others. Elementary reactions follow the law of mass action, but the rate law of stepwise reactions has to be derived by combining the rate laws of the various elementary steps, and can become rather complex. In consecutive reactions, the rate-determining step often determines the kinetics. In consecutive first order reactions, a steady state approximation can simplify the rate law. The activation energy for a reaction is experimentally determined through the Arrhenius equation and the Eyring equation. The main factors that influence the reaction rate include: the physical state of the reactants, the concentrations of the reactants, the temperature at which the reaction occurs, and whether or not any catalysts are present in the reaction.

Gorban and Yablonsky have suggested that the history of chemical dynamics can be divided into three eras. The first is the van 't Hoff wave searching for the general laws of chemical reactions and relating kinetics to thermodynamics. The second may be called the Semenov--Hinshelwood wave with emphasis on reaction mechanisms, especially for chain reactions. The third is associated with Aris and the detailed mathematical description of chemical reaction networks.

Factors Affecting Reaction Rate

Nature of the Reactants

Depending upon what substances are reacting, the reaction rate varies. Acid/base reactions, the formation of salts, and ion exchange are fast reactions. When covalent bond formation takes place between the molecules and when large molecules are formed, the reactions tend to be very slow. Nature and strength of bonds in reactant molecules greatly influence the rate of its transformation into products.

Physical State

The physical state (solid, liquid, or gas) of a reactant is also an important factor of the rate of change. When reactants are in the same phase, as in aqueous solution, thermal motion brings them into contact. However, when they are in different phases, the reaction is limited to the interface between the reactants. Reaction can occur only at their area of contact; in the case of a liquid and a gas, at the surface of the liquid. Vigorous shaking and stirring may be needed to bring the reaction to completion. This means that the more finely divided a solid or liquid reactant the greater its surface area per unit volume and the more contact it with the other reactant, thus the faster the reaction. To make an analogy, for example, when one starts a fire, one uses wood chips and small branches — one does not start with large logs right away. In organic chemistry, on water reactions are the exception to the rule that homogeneous reactions take place faster than heterogeneous reactions.

Surface Area of Solids

In a solid, only those particles that are at the surface can be involved in a reaction. Crushing a solid into smaller parts means that more particles are present at the surface, and the frequency of collisions between these and reactant particles increases, and so reaction occurs more rapidly. For example, Sherbet (powder) is a mixture of very fine powder of malic acid (a weak organic acid) and sodium hydrogen carbonate. On contact with the saliva in the mouth, these chemicals quickly dissolve and react, releasing carbon dioxide and providing for the fizzy sensation. Also, fireworks manufacturers modify the surface area of solid reactants to control the rate at which the fuels in fireworks are oxidised, using this to create different effects. For example, finely divided aluminium confined in a shell explodes violently. If larger pieces of aluminium are used, the reaction is slower and sparks are seen as pieces of burning metal are ejected.

Concentration

The reactions are due to collisions of reactant species. The frequency with which the molecules or ions collide depends upon their concentrations. The more crowded the molecules are, the more likely they are to collide and react with one another. Thus, an increase in the concentrations of the reactants will usually result in the corresponding increase in the reaction rate, while a decrease in the concentrations will usually have a reverse effect. For example, combustion will occur more rapidly in pure oxygen than in air (21% oxygen).

The rate equation shows the detailed dependence of the reaction rate on the concentrations of reactants and other species present. Different mathematical forms are possible depending on the reaction mechanism. The actual rate equation for a given reaction is determined experimentally and provides information about the reaction mechanism.

Temperature

Temperature usually has a major effect on the rate of a chemical reaction. Molecules at a higher temperature have more thermal energy. Although collision frequency is greater at higher temperatures, this alone contributes only a very small proportion to the increase in rate of reaction. Much more important is the fact that the proportion of reactant molecules with sufficient energy to react (energy greater than activation energy: $E > E_a$) is significantly higher and is explained in detail by the Maxwell–Boltzmann distribution of molecular energies.

The 'rule of thumb' that the rate of chemical reactions doubles for every 10 °C temperature rise is a common misconception. This may have been generalized from the special case of biological systems, where the α (temperature coefficient) is often between 1.5 and 2.5.

A reaction's kinetics can also be studied with a temperature jump approach. This involves using a sharp rise in temperature and observing the relaxation time of the return to equilibrium. A particularly useful form of temperature jump apparatus is a shock tube, which can rapidly jump a gas's temperature by more than 1000 degrees.

Catalysts

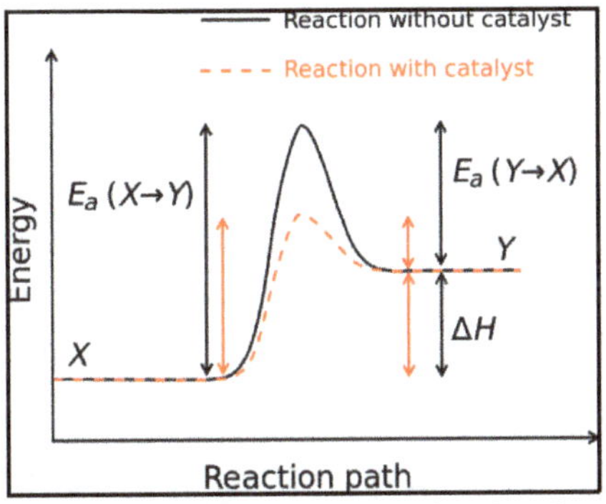

Generic potential energy diagram showing the effect of a catalyst in a hypothetical endothermic chemical reaction. The presence of the catalyst opens a different reaction pathway (shown in red) with a lower activation energy. The final result and the overall thermodynamics are the same.

A catalyst is a substance that alters the rate of a chemical reaction but remains chemically unchanged afterwards. The catalyst increases the rate of the reaction by providing a different reaction mechanism to occur with a lower activation energy. In autocatalysis a reaction product is itself a catalyst for that reaction leading to positive feedback. Proteins that act as catalysts in biochemical reactions are called enzymes. Michaelis–Menten kinetics describe the rate of enzyme mediated reactions. A catalyst does not affect the position of the equilibrium, as the catalyst speeds up the backward and forward reactions equally.

In certain organic molecules, specific substituents can have an influence on reaction rate in neighbouring group participation.

Pressure

Increasing the pressure in a gaseous reaction will increase the number of collisions between reactants, increasing the rate of reaction. This is because the activity of a gas is directly proportional to the partial pressure of the gas. This is similar to the effect of increasing the concentration of a solution.

In addition to this straightforward mass-action effect, the rate coefficients themselves can change due to pressure. The rate coefficients and products of many high-temperature gas-phase reactions change if an inert gas is added to the mixture; variations on this effect are called fall-off and chemical activation. These phenomena are due to exothermic or endothermic reactions occurring faster than heat transfer, causing the reacting molecules to have non-thermal energy distributions (non-Boltzmann distribution). Increasing the pressure increases the heat transfer rate between the reacting molecules and the rest of the system, reducing this effect.

Condensed-phase rate coefficients can also be affected by (very high) pressure; this is a completely different effect than fall-off or chemical-activation. It is often studied using diamond anvils.

A reaction's kinetics can also be studied with a pressure jump approach. This involves making fast changes in pressure and observing the relaxation time of the return to equilibrium.

Experimental Methods

The experimental determination of reaction rates involves measuring how the concentrations of reactants or products change over time. For example, the concentration of a reactant can be measured by spectrophotometry at a wavelength where no other reactant or product in the system absorbs light.

For reactions which take at least several minutes, it is possible to start the observations after the reactants have been mixed at the temperature of interest.

Fast Reactions

For faster reactions, the time required to mix the reactants and bring them to a specified temperature may be comparable or longer than the half-life of the reaction. Special methods to start fast reactions without slow mixing step include

- Stopped flow methods, which can reduce the mixing time to the order of a millisecond.
- Chemical relaxation methods such as temperature jump and pressure jump, in which a pre-mixed system initially at equilibrium is perturbed by rapid heating or depressurization so that it is no longer at equilibrium, and the relaxation back to equilibrium is observed. For example, this method has been used to study the neutralization $H_3O^+ + OH^-$ with a half-life of 1 μs or less under ordinary conditions.
- Flash photolysis, in which a laser pulse produces highly excited species such as free radicals, whose reactions are then studied.

Equilibrium

While chemical kinetics is concerned with the rate of a chemical reaction, thermodynamics determines the extent to which reactions occur. In a reversible reaction, chemical equilibrium is reached when the rates of the forward and reverse reactions are equal (the principle of detailed balance) and the concentrations of the reactants and products no longer change. This is demonstrated by, for example, the Haber–Bosch process for combining nitrogen and hydrogen to produce ammonia. Chemical clock reactions such as the Belousov–Zhabotinsky reaction demonstrate that component concentrations can oscillate for a long time before finally attaining the equilibrium.

Free Energy

In general terms, the free energy change (ΔG) of a reaction determines whether a chemical change will take place, but kinetics describes how fast the reaction is. A reaction can be very exothermic and have a very positive entropy change but will not happen in practice if the reaction is too slow. If a reactant can produce two different products, the thermodynamically most stable one will in general form, except in special circumstances when the reaction is said to be under kinetic reaction control. The Curtin–Hammett principle applies when determining the product ratio for two reactants interconverting rapidly, each going to a different product. It is possible to make predictions about reaction rate constants for a reaction from free-energy relationships.

The kinetic isotope effect is the difference in the rate of a chemical reaction when an atom in one of the reactants is replaced by one of its isotopes.

Chemical kinetics provides information on residence time and heat transfer in a chemical reactor in chemical engineering and the molar mass distribution in polymer chemistry.

Applications and Models

The mathematical models that describe chemical reaction kinetics provide chemists and chemical engineers with tools to better understand and describe chemical processes such as food decomposition, microorganism growth, stratospheric ozone decomposition, and the complex chemistry of biological systems. These models can also be used in the design or modification of chemical reactors to optimize product yield, more efficiently separate products, and eliminate environmentally harmful by-products. When performing catalytic cracking of heavy hydrocarbons into gasoline and light gas, for example, kinetic models can be used to find the temperature and pressure at which the highest yield of heavy hydrocarbons into gasoline will occur.

Chemical Kinetics is frequently validated and explored through modeling in specialized packages as a function of ordinary differential equation-solving (ODE-solving) and curve-fitting.

References

- Lienhard, John H. (2011). A Heat Transfer Textbook. Dover Civil and Mechanical Engineering Series. Courier Corporation. ISBN 9780486479316.
- Plawsky, Joel L. (April 2001). Transport phenomena fundamentals (Chemical Industries Series). CRC Press. pp. 1, 2, 3. ISBN 978-0-8247-0500-8.
- Truskey, George; Yuan F; Katz D. Transport Phenomena in Biological Systems (Second ed.). Prentice Hall. p. 888. ISBN 978-0131569881.

- Ott, Bevan J.; Boerio-Goates, Juliana (2000). Chemical Thermodynamics – Principles and Applications. Academic Press. ISBN 0-12-530990-2.
- Delgass; et al. "Seventy Five Years of Chemical Engineering". Prudue University. Retrieved 01, April 2020
- "Year in Science, 2010", Interview with Secretary of Energy, Steven Chu, Discover magazine, Jan-Feb 2011, p. 42.
- A.N. Gorban, G.S. Yablonsky Three Waves of Chemical Dynamics, Mathematical Modelling of Natural Phenomena 10(5) (2015), p. 1–5.
- Espenson, J.H. Chemical Kinetics and Reaction Mechanisms (2nd ed., McGraw-Hill 2002), p.264-6 ISBN 0-07-288362-6

5

Chemical Engineering: Areas of Applications

Chemical engineering is used in a wide range of fields and industries such as food engineering, plastics engineering, pharmaceutical engineering, tissue engineering, chemometrics and ceramic engineering. The topics elaborated in this chapter will help in gaining a better perspective about these applications of chemical engineering.

Food Engineering

Bread factory in Germany.

Food engineering is a multidisciplinary field of applied physical sciences which combines science, microbiology, and engineering education for food and related industries. Food engineering includes, but is not limited to, the application of agricultural engineering, mechanical engineering and chemical engineering principles to food materials. Food engineers provide the technological knowledge transfer essential to the cost-effective production and commercialization of food products and services. Physics, chemistry, and mathematics are fundamental to understanding and engineering products and operations in the food industry.

Food engineering encompasses a wide range of activities. Food engineers are employed in food processing, food machinery, packaging, ingredient manufacturing, instrumentation, and control. Firms that design and build food processing plants, consulting firms, government agencies, pharmaceutical companies, and health-care firms also employ food engineers. Specific food engineering activities include:

- Drug/food products;
- Design and installation of food/biological/pharmaceutical production processes;
- Design and operation of environmentally responsible waste treatment systems;
- Marketing and technical support for manufacturing plants.

Topics in Food Engineering

In the development of food engineering, one of the many challenges is to employ modern tools, technology, and knowledge, such as computational materials science and nanotechnology, to develop new products and processes. Simultaneously, improving quality, safety, and security remain critical issues in food engineering study. New packaging materials and techniques are being developed to provide more protection to foods, and novel preservation technology is emerging. Additionally, process control and automation regularly appear among the top priorities identified in food engineering. Advanced monitoring and control systems are developed to facilitate automation and flexible food manufacturing. Furthermore, energy saving and minimization of environmental problems continue to be important food engineering issues, and significant progress is being made in waste management, efficient utilization of energy, and reduction of effluents and emissions in food production.

Typical topics include:food people

- Advances in classical unit operations in engineering applied to food manufacturing.
- Progresses in the transport and storage of liquid and solid foods.
- Developments in heating, chilling and freezing of foods.
- Advanced mass transfer in foods.
- New chemical and biochemical aspects of food engineering and the use of kinetic analysis.
- New techniques in dehydration, thermal processing, non-thermal processing, extrusion, liquid food concentration, membrane processes and applications of membranes in food processing.
- Shelf-life, electronic indicators in inventory management, and sustainable technologies in food processing.

- Modern packaging, cleaning, and sanitation technologies.
- Development of sensors systems for quality and safety assessment.

Plastics Engineering

Plastics engineering encompasses the processing, design, development, and manufacture of plastics products. A plastic is a polymeric material that is in a semi-liquid state, having the property of plasticity and exhibiting flow. Plastics engineering encompasses plastics material and plastic machinery. Plastic Machinery is the general term for all types of machinery and devices used in the plastics processing industry. The nature of plastic materials poses unique challenges to an engineer. Mechanical properties of plastics are often difficult to quantify, and the plastics engineer has to design a product that meets certain specifications while keeping costs to a minimum. Other properties that the plastics engineer has to address include: outdoor weatherability, thermal properties such as upper use temperature, electrical properties, barrier properties, and resistance to chemical attack.

In plastics engineering, as in most engineering disciplines, the economics of a product plays an important role. The cost of plastic materials ranges from the cheapest commodity plastics used in mass-produced consumer products to the very expensive, specialty plastics. The cost of a plastic product is measured in different ways, and the absolute cost of a plastic material is difficult to ascertain. Cost is often measured in price per pound of material, or price per unit volume of material. In many cases however, it is important for a product to meet certain specifications, and cost could then be measured in price per unit of a property. Price with respect to processibility is often important, as some materials need to be processed at very high temperatures, increasing the amount of cooling time a part needs. In a large production run cooling time is very expensive.

Some plastics are manufactured from re-cycled materials but their use in engineering tends to be limited because the consistency of formulation and their physical properties tend to be less consistent. Electrical and electronic equipment and motor vehicle markets together accounted for 58 percent of engineered plastics demand in 2003. Engineered plastics demand in the US was estimated at $9,702 million in 2007.

A big challenge for plastics engineers is the reduction of the ecological footprints of their products. First attempts like the Vinyloop process can guarantee that a product's primary energy demand is 46 percent lower than conventional produced PVC. The global warming potential is 39 percent lower.

Plastics Engineering Specialties

- Consumer Plastics.
- Medical plastics.

- Automotive plastics.
- Recycled or recyclable plastics.
- Biodegradable plastics.
- Elastomers / rubber.
- Epoxys.
- Plastics processing: injection moulding, plastics extrusion, stretch-blow molding, thermoforming, compression molding, calendering, transfer molding, laminating, fiberglass molding, pultrusion, filament winding, vacuum forming, rotational molding.
- Ultrasonic welding.

Pharmaceutical Engineering

Pharmaceutical engineering is a branch of pharmaceutical science and technology that involves development and manufacturing of products, processes, and components in the pharmaceuticals industry (i.e. drugs & biologics). While developing pharmaceutical products involves many interrelated disciplines (e.g. medicinal chemists, analytical chemists, clinicians/pharmacologists, pharmacists, chemical engineers, biomedical engineers, etc.), the specific subfield of "pharmaceutical engineering" has only emerged recently as a distinct engineering discipline. This now brings the problem-solving principles and quantitative training of engineering to complement the other scientific fields already involved in drug development.

Academic Programs

There are still relatively few academic programs with this explicit focus. The first one began at the University of Michigan, as a joint project between their College of Engineering and School of Pharmacy. Because such programs are not yet common, many pharmaceutical engineers have had their formal engineering training in chemical or biomedical engineering.

Most pharmaceutical engineering programs are graduate-level, and as with biomedical engineering there is generally an expectation that engineers and scientists working in pharmaceutical engineering should have some relevant graduate-level education. Many have a masters or PhD degree in chemical or biomedical engineering, or a related science. In Italy there is a university degree course (5 years) in Chemistry and Pharmaceutical Technologies (Chimica e Tecnologie Farmaceutiche), different from pharmacy, that ability as a pharmacist and different roles in the industry as an engineer (for the Italian legislation is not really an engineer though performs the same tasks).

Professional Licensure and Certification

In most jurisdictions, engineering licensure (e.g. Licensed "Professional Engineer" or P.E.) is not discipline-specific, so any licensed engineer with competency in pharmaceutical engineering may qualify as licensed. However, in the U.S., most pharmaceutical engineers fall under the "industrial exemption," which does not require a (P.E.) license for those engineers whose work is completely internal and for a private employer. There are ongoing debates about whether to narrow or eliminate this exemption from engineering licensure, and the Executive Director of the National Society of Professional Engineers (NSPE) recently advocated requiring licensure for engineers in the pharmaceutical industry (among a few others).

Most U.S. jurisdictions require two examinations as part of their licensing criteria, the second of which allows electing a particular discipline of emphasis (while not affecting the license itself, as noted above), so pharmaceutical engineers are likely to select the testing option of either chemical engineering or biomedical engineering (which are generally considered the closest available options).

There is also a private (non-governmental) certification offered by the professional organization International Society for Pharmaceutical Engineering (ISPE), known as Certified Pharmaceutical Industry Professional (CPIP). This tends to focus more on (later-stage) manufacturing and commercialization issues, etc., rather than early-stage things like drug design, discovery assays, and preclinical development.

Common Specialties

- Pharmaceutical Development Sciences - broadly, assays or techniques for discovering, modifying, or designing drug substances or excipients; in particular, rational drug design - as a relatively recent alternative to traditional trial-and-error drug discovery processes - relies upon principles of engineering more than many other pharmaceutical sciences such as formulation or medicinal chemistry.
- Bio-/Pharmaceutical Manufacturing Science - optimal processes for producing drug substances & products with quality and efficiency.
- Clinical Science - applying engineering principles toward conduct of studies to assess safety & efficacy, for the medical community and regulators.
- Regulatory Science - scientific bases for regulatory decision-making (typically by the FDA, in the U.S.), with an emphasis on risk-benefit analysis.
- Pharmaceutical Devices - designing instruments, tools, or implants which facilitate the making, handling, or use of drugs (e.g. drug delivery chips).

These specialties overlap with other engineering areas as well as non-engineering scientific and medical fields, although in all specialties Pharmaceutical Engineers tend to have a distinct focus on product and process design and quantitative analysis. And in addition to these technical areas, some pharmaceutical engineers pursue careers as business or legal professionals. Their scientific and engineering background is often suitable for careers in management, patent law, or even entrepreneurship - for example.

Tissue Engineering

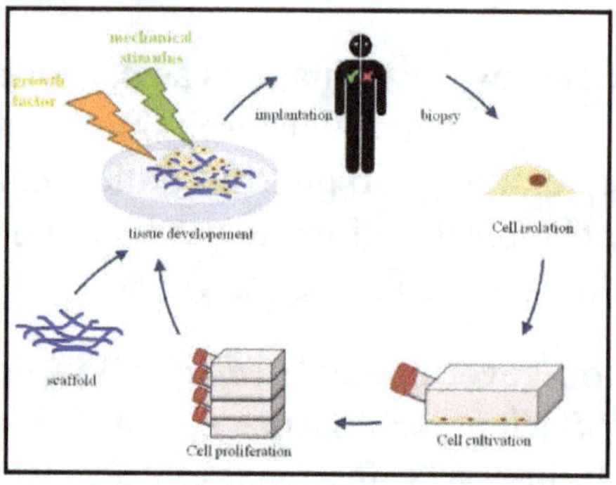

Principle of tissue engineering.

Tissue engineering is the use of a combination of cells, engineering and materials methods, and suitable biochemical and physicochemical factors to improve or replace biological tissues. Tissue engineering involves the use of a scaffold for the formation of new viable tissue for a medical purpose. While it was once categorized as a sub-field of biomaterials, having grown in scope and importance it can be considered as a field in its own.

While most definitions of tissue engineering cover a broad range of applications, in practice the term is closely associated with applications that repair or replace portions of or whole tissues (i.e., bone, cartilage, blood vessels, bladder, skin, muscle etc.). Often, the tissues involved require certain mechanical and structural properties for proper functioning. The term has also been applied to efforts to perform specific biochemical functions using cells within an artificially-created support system (e.g. an artificial pancreas, or a bio artificial liver). The term regenerative medicine is often used synonymously with tissue engineering, although those involved in regenerative medicine place more emphasis on the use of stem cells or progenitor cells to produce tissues.

Overview

A commonly applied definition of tissue engineering, as stated by Langer and Vacanti, is "an interdisciplinary field that applies the principles of engineering and life sciences toward the development of biological substitutes that restore, maintain, or improve

[Biological tissue] function or a whole organ". Tissue engineering has also been defined as "understanding the principles of tissue growth, and applying this to produce functional replacement tissue for clinical use." A further description goes on to say that an "underlying supposition of tissue engineering is that the employment of natural biology of the system will allow for greater success in developing therapeutic strategies aimed at the replacement, repair, maintenance, and/or enhancement of tissue function."

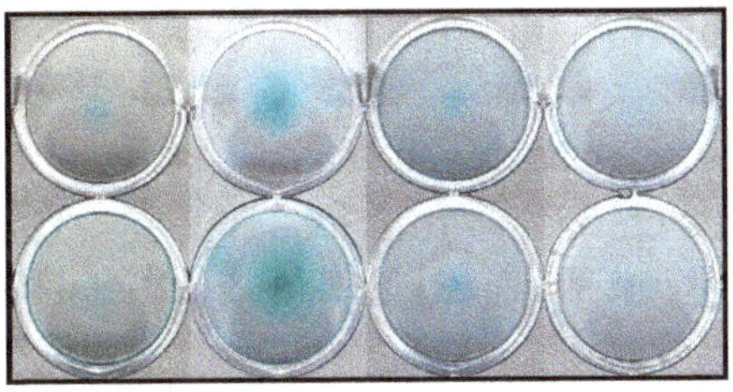

Micro-mass cultures of C3H-10T1/2 cells at varied oxygen tensions stained with Alcian blue.

Powerful developments in the multidisciplinary field of tissue engineering have yielded a novel set of tissue replacement parts and implementation strategies. Scientific advances in biomaterials, stem cells, growth and differentiation factors, and biomimetic environments have created unique opportunities to fabricate tissues in the laboratory from combinations of engineered extracellular matrices ("scaffolds"), cells, and biologically active molecules. Among the major challenges now facing tissue engineering is the need for more complex functionality, as well as both functional and biomechanical stability and vascularization in laboratory-grown tissues destined for transplantation. The continued success of tissue engineering, and the eventual development of true human replacement parts, will grow from the convergence of engineering and basic research advances in tissue, matrix, growth factor, stem cell, and developmental biology, as well as materials science and bio informatics.

In 2003, the NSF published a report entitled "The Emergence of Tissue Engineering as a Research Field", which gives a thorough description of the history of this field.

Examples

- Bioartificial windpipe: The first procedure of regenerative medicine of an implantation of a "bioartificial" organ.
- In vitro meat: Edible artificial animal muscle tissue cultured in vitro.
- Bioartificial liver device: several research efforts have produced hepatic assist devices utilizing living hepatocytes.
- Artificial pancreas: research involves using islet cells to produce and regulate insulin, particularly in cases of diabetes.
- Artificial bladders: Anthony Atala (Wake Forest University) has successfully implanted artificially grown bladders into seven out of approximately 20 human test subjects as part of a long-term experiment.

- Cartilage: lab-grown tissue was successfully used to repair knee cartilage.
- Scaffold-free cartilage: Cartilage generated without the use of exogenous scaffold material. In this methodology, all material in the construct is cellular or material produced directly by the cells themselves.
- Doris Taylor's heart in a jar.
- Tissue-engineered airway.
- Tissue-engineered vessels.
- Artificial skin constructed from human skin cells embedded in a hydrogel, such as in the case of bioprinted constructs for battlefield burn repairs.
- Artificial bone marrow.
- Artificial bone.
- Laboratory-grown penis.
- Oral mucosa tissue engineering.
- Foreskin.

Cells as Building Blocks

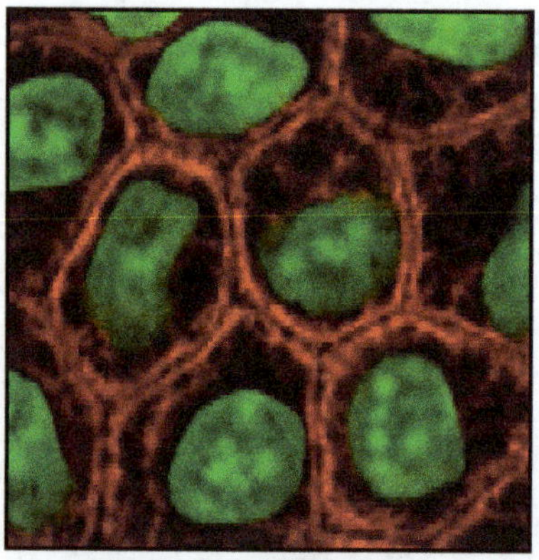

Stained cells in culture.

Tissue engineering utilizes living cells as engineering materials. Examples include using living fibroblasts in skin replacement or repair, cartilage repaired with living chondrocytes, or other types of cells used in other ways.

Cells became available as engineering materials when scientists at Geron Corp. discovered how to extend telomeres in 1998, producing immortalized cell lines. Before this, laboratory cultures of healthy, noncancerous mammalian cells would only divide a fixed number of times, up to the Hayflick limit, before dying.

Extraction

From fluid tissues such as blood, cells are extracted by bulk methods, usually centrifugation or apheresis. From solid tissues, extraction is more difficult. Usually the tissue is minced, and then digested with the enzymes trypsin or collagenase to remove the extracellular matrix (ECM) that holds the cells. After that, the cells are free floating, and extracted using centrifugation or apheresis.Digestion with trypsin is very dependent on temperature. Higher temperatures digest the matrix faster, but create more damage. Collagenase is less temperature dependent, and damages fewer cells, but takes longer and is a more expensive reagent.

Types of Cells

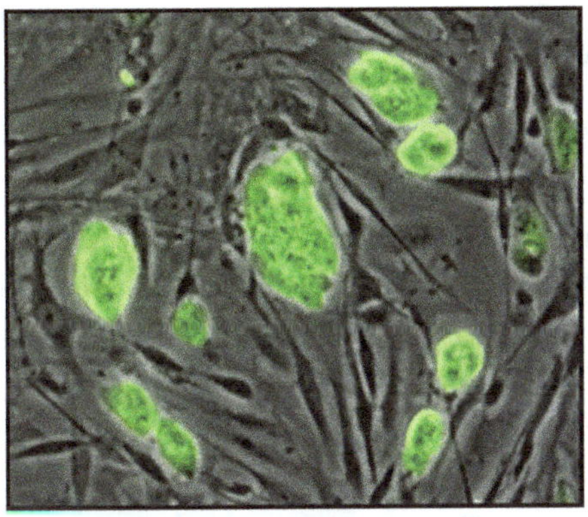

Mouse embryonic stem cells.

Cells are often categorized by their source:

- Autologous cells are obtained from the same individual to which they will be re-implanted. Autologous cells have the fewest problems with rejection and pathogen transmission, however in some cases might not be available. For example, in genetic disease suitable autologous cells are not available. Also very ill or elderly persons, as well as patients suffering from severe burns, may not have sufficient quantities of autologous cells to establish useful cell lines. Moreover, since this category of cells needs to be harvested from the patient, there are also some concerns related to the necessity of performing such surgical operations that might lead to donor site infection or chronic pain. Autologous cells also must be cultured from samples before they can be used: this takes time, so autologous solutions may not be very quick. Recently there has been a trend towards the use of mesenchymal stem cells from bone marrow and fat. These cells can differentiate into a variety of tissue types, including bone, cartilage, fat, and nerve. A large number of cells can be easily and quickly isolated from fat, thus opening the potential for large numbers of cells to be quickly and easily obtained.
- Allogeneic cells come from the body of a donor of the same species. While there

are some ethical constraints to the use of human cells for in vitro studies, the employment of dermal fibroblasts from human foreskin has been demonstrated to be immunologically safe and thus a viable choice for tissue engineering of skin.

- Xenogenic cells are these isolated from individuals of another species. In particular animal cells have been used quite extensively in experiments aimed at the construction of cardiovascular implants.
- Syngenic or isogenic cells are isolated from genetically identical organisms, such as twins, clones, or highly inbred research animal models.
- Primary cells are from an organism.
- Secondary cells are from a cell bank.
- Stem cells are undifferentiated cells with the ability to divide in culture and give rise to different forms of specialized cells. According to their source stem cells are divided into "adult" and "embryonic" stem cells, the first class being multipotent and the latter mostly pluripotent; some cells are totipotent, in the earliest stages of the embryo. While there is still a large ethical debate related with the use of embryonic stem cells, it is thought that another alternative source - induced stem cells may be useful for the repair of diseased or damaged tissues, or may be used to grow new organs.

Scaffolds

Scaffolds are materials that have been engineered to cause desirable cellular interactions to contribute to the formation of new functional tissues for medical purposes. Cells are often 'seeded' into these structures capable of supporting three-dimensional tissue formation. Scaffolds mimic the extracellular matrix of the native tissue, recapitulating the in vivo milieu and allowing cells to influence their own microenvironments. They usually serve for at least one of the following purposes:

- Allow cell attachment and migration.
- Deliver and retain cells and biochemical factors.
- Enable diffusion of vital cell nutrients and expressed products.
- Exert certain mechanical and biological influences to modify the behaviour of the cell phase.

In 2009, an interdisciplinary team led by the thoracic surgeon Thorsten Walles implanted the first bioartificial transplant that provides an innate vascular network for post-transplant graft supply successfully into a patient awaiting tracheal reconstruction.

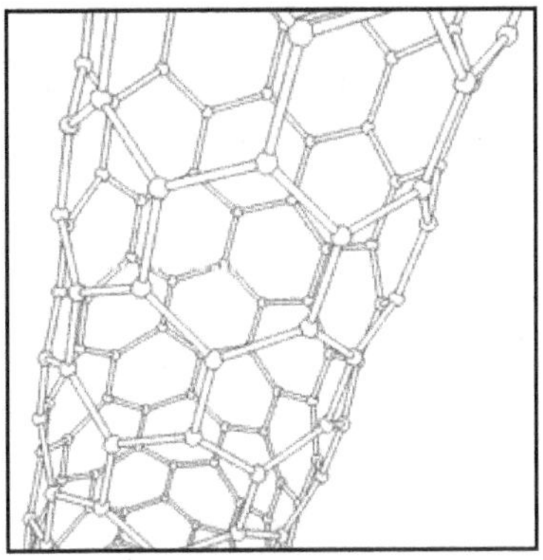

This animation of a rotating carbon nanotube shows its 3D structure. Carbon nanotubes are among the numerous candidates for tissue engineering scaffolds since they are biocompatible, resistant to biodegradation and can be functionalized with biomolecules. However, the possibility of toxicity with non-biodegradable nano-materials is not fully understood.

To achieve the goal of tissue reconstruction, scaffolds must meet some specific requirements. A high porosity and an adequate pore size are necessary to facilitate cell seeding and diffusion throughout the whole structure of both cells and nutrients. Biodegradability is often an essential factor since scaffolds should preferably be absorbed by the surrounding tissues without the necessity of a surgical removal. The rate at which degradation occurs has to coincide as much as possible with the rate of tissue formation: this means that while cells are fabricating their own natural matrix structure around themselves, the scaffold is able to provide structural integrity within the body and eventually it will break down leaving the newly formed tissue which will take over the mechanical load. Injectability is also important for clinical uses. Recent research on organ printing is showing how crucial a good control of the 3D environment is to ensure reproducibility of experiments and offer better results.

Materials

Many different materials (natural and synthetic, biodegradable and permanent) have been investigated. Most of these materials have been known in the medical field before the advent of tissue engineering as a research topic, being already employed as bioresorbable sutures. Examples of these materials are collagen and some polyesters.

New biomaterials have been engineered to have ideal properties and functional customization: injectability, synthetic manufacture, biocompatibility, non-immunogenicity, transparency, nano-scale fibers, low concentration, resorption rates, etc. PuraMatrix, originating from the MIT labs of Zhang, Rich, Grodzinsky and Langer is one of these new biomimetic scaffold families which has now been commercialized and is impacting clinical tissue engineering.

A commonly used synthetic material is PLA - polylactic acid. This is a polyester which degrades within the human body to form lactic acid, a naturally occurring chemical which is easily removed from the body. Similar materials are polyglycolic acid (PGA) and polycaprolactone (PCL): their degradation mechanism is similar to that of PLA, but they exhibit respectively a faster and a slower rate of degradation compared to PLA. While these materials have well maintained mechanical strength and structural integrity, they exhibit a hydrophobic nature. This hydrophobicity inhibits their biocompatibility, which makes them less effective for in vivo use as tissue scaffolding. In order to fix the lack of biocompatibility, much research has been done to combine these hydrophobic materials with hydrophilic and more biocompatible hydrogels. While these hydrogels have a superior biocompatibility, they lack the structural integrity of PLA, PCL, and PGA. By combining the two different types of materials, researchers are trying to create a synergistic relationship that produces a more biocompatible tissue scaffolding. Scaffolds may also be constructed from natural materials: in particular different derivatives of the extracellular matrix have been studied to evaluate their ability to support cell growth. Proteic materials, such as collagen or fibrin, and polysaccharidic materials, like chitosan or glycosaminoglycans (GAGs), have all proved suitable in terms of cell compatibility, but some issues with potential immunogenicity still remains. Among GAGs hyaluronic acid, possibly in combination with cross linking agents (e.g. glutaraldehyde, water-soluble carbodiimide, etc.), is one of the possible choices as scaffold material. Functionalized groups of scaffolds may be useful in the delivery of small molecules (drugs) to specific tissues. Another form of scaffold under investigation is decellularised tissue extracts whereby the remaining cellular remnants/extracellular matrices act as the scaffold. Recently a range of nanocomposites biomaterials are fabricated by incorporating nanomaterials within polymeric matrix to engineer bioactive scaffolds.

A 2009 study by Ratmir et al. aimed to improve in vivo-like conditions for 3D tissue via "stacking and de-stacking layers of paper impregnated with suspensions of cells in extracellular matrix hydrogel, making it possible to control oxygen and nutrient gradients in 3D, and to analyze molecular and genetic responses". It is possible to manipulate gradients of soluble molecules, and to characterize cells in these complex gradients more effectively than conventional 3D cultures based on hydrogels, cell spheroids, or 3D perfusion reactors. Different thicknesses of paper and types of medium can support a variety of experimental environments. Upon deconstruction, these sheets can be useful in cell-based high-throughput screening and drug discovery.

Synthesis

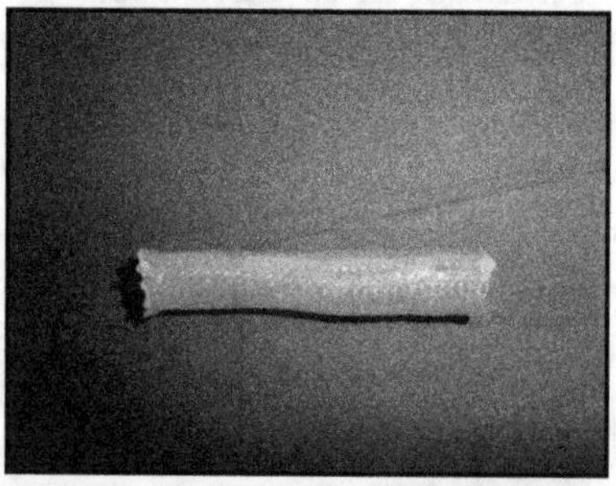

Tissue engineered vascular graft.

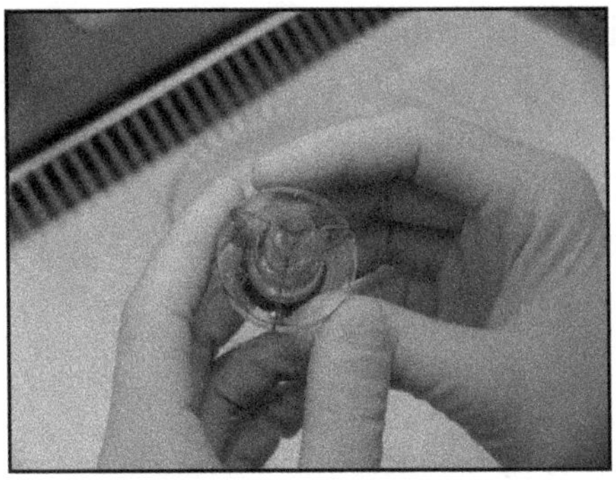

Tissue engineered heart valve.

A number of different methods have been described in literature for preparing porous structures to be employed as tissue engineering scaffolds. Each of these techniques presents its own advantages, but none are free of drawbacks.

Nanofiber self-assembly:

> Molecular self-assembly is one of the few methods for creating biomaterials with properties similar in scale and chemistry to that of the natural in vivo extracellular matrix (ECM), a crucial step toward tissue engineering of complex tissues. Moreover, these hydrogel scaffolds have shown superiority in in vivo toxicology and biocompatibility compared to traditional macroscaffolds and animal-derived materials.

Textile technologies:

> These techniques include all the approaches that have been successfully employed for the preparation of non-woven meshes of different polymers. In particular, non-woven polyglycolide structures have been tested for tissue engineering applications: such fibrous structures have been found useful to grow different types of cells. The principal drawbacks are related to the difficulties in obtaining high porosity and regular pore size.

Solvent casting and particulate leaching (SCPL):

> This approach allows for the preparation of structures with regular porosity, but with limited thickness. First, the polymer is dissolved into a suitable organic solvent (e.g. polylactic acid could be dissolved into dichloromethane), then the solution is cast into a mold filled with porogen particles. Such porogen can be an inorganic salt like sodium chloride, crystals of saccharose, gelatin spheres or paraffin spheres. The size of the porogen particles will affect the size of the scaffold pores, while the polymer to porogen ratio is directly correlated to the amount of porosity of the final structure. After the polymer solution has been cast the solvent is allowed to fully evaporate, then the composite structure in the mold is immersed in a bath of a liquid suitable for dissolving the porogen: water

in the case of sodium chloride, saccharose and gelatin or an aliphatic solvent like hexane for use with paraffin. Once the porogen has been fully dissolved, a porous structure is obtained. Other than the small thickness range that can be obtained, another drawback of SCPL lies in its use of organic solvents which must be fully removed to avoid any possible damage to the cells seeded on the scaffold.

Gas foaming:

To overcome the need to use organic solvents and solid porogens, a technique using gas as a porogen has been developed. First, disc-shaped structures made of the desired polymer are prepared by means of compression molding using a heated mold. The discs are then placed in a chamber where they are exposed to high pressure CO_2 for several days. The pressure inside the chamber is gradually restored to atmospheric levels. During this procedure the pores are formed by the carbon dioxide molecules that abandon the polymer, resulting in a sponge-like structure. The main problems resulting from such a technique are caused by the excessive heat used during compression molding (which prohibits the incorporation of any temperature labile material into the polymer matrix) and by the fact that the pores do not form an interconnected structure.

Emulsification/Freeze-drying:

This technique does not require the use of a solid porogen like SCPL. First, a synthetic polymer is dissolved into a suitable solvent (e.g. polylactic acid in dichloromethane) then water is added to the polymeric solution and the two liquids are mixed in order to obtain an emulsion. Before the two phases can separate, the emulsion is cast into a mold and quickly frozen by means of immersion into liquid nitrogen. The frozen emulsion is subsequently freeze-dried to remove the dispersed water and the solvent, thus leaving a solidified, porous polymeric structure. While emulsification and freeze-drying allow for a faster preparation when compared to SCPL (since it does not require a time consuming leaching step), it still requires the use of solvents. Moreover, pore size is relatively small and porosity is often irregular. Freeze-drying by itself is also a commonly employed technique for the fabrication of scaffolds. In particular, it is used to prepare collagen sponges: collagen is dissolved into acidic solutions of acetic acid or hydrochloric acid that are cast into a mold, frozen with liquid nitrogen and then lyophilized.

Thermally induced phase separation (TIPS):

Similar to the previous technique, this phase separation procedure requires the use of a solvent with a low melting point that is easy to sublime. For example, dioxane could be used to dissolve polylactic acid, then phase separation is induced through the addition of a small quantity of water: a polymer-rich and a polymer-poor phase are formed. Following cooling below the solvent melting point and some days of vacuum-drying to sublime the solvent, a porous scaffold is obtained. Liquid-liquid phase separation presents the same drawbacks of emulsification/freeze-drying.

Electrospinning:

A highly versatile technique that can be used to produce continuous fibers from submicrometer to nanometer diameters. In a typical electrospinning set-up, a solution is fed through a spinneret and a high voltage is applied to the tip. The buildup of electrostatic repulsion within the charged solution, causes it to eject a thin fibrous stream. A mounted collector plate or rod with an opposite or grounded charge draws in the continuous fibers, which arrive to form a highly porous network. The primary advantages of this technique are its simplicity and ease of variation. At a laboratory level, a typical electrospinning set-up only requires a high voltage power supply (up to 30 kV), a syringe, a flat tip needle and a conducting collector. For these reasons, electrospinning has become a common method of scaffold manufacture in many labs. By modifying variables such as the distance to collector, magnitude of applied voltage, or solution flow rate—researchers can dramatically change the overall scaffold architecture.

CAD/CAM technologies:

Because most of the above techniques are limited when it comes to the control of porosity and pore size, computer assisted design and manufacturing techniques have been introduced to tissue engineering. First, a three-dimensional structure is designed using CAD software. The porosity can be tailored using algorithms within the software. The scaffold is then realized by using ink-jet printing of polymer powders or through Fused Deposition Modeling of a polymer melt.

A 2011 study by El-Ayoubi et al. investigated "3D-plotting technique to produce (biocompatible and biodegradable) poly-L-Lactide macroporous scaffolds with two different pore sizes" via solid free-form fabrication (SSF) with computer-aided-design (CAD), to explore therapeutic articular cartilage replacement as an "alternative to conventional tissue repair". The study found the smaller the pore size paired with mechanical stress in a bioreactor (to induce in vivo-like conditions), the higher the cell viability in potential therapeutic functionality via decreasing recovery time and increasing transplant effectiveness.

Laser-assisted BioPrinting (LaBP):

In a 2012 study, Koch et al. focused on whether Laser-assisted BioPrinting (LaBP) can be used to build multicellular 3D patterns in natural matrix, and whether the generated constructs are functioning and forming tissue. LaBP arranges small volumes of living cell suspensions in set high-resolution patterns.

The investigation was successful, the researchers foresee that "generated tissue constructs might be used for in vivo testing by implanting them into animal models" (14). As of this study, only human skin tissue has been synthesized, though researchers project that "by integrating further cell types (e.g. melanocytes, Schwann cells, hair follicle cells) into the printed cell construct, the behavior of these cells in a 3D in vitro microenvironment similar to their natural one can be analyzed", useful for drug discovery and toxicology studies.

Assembly Methods

One of the continuing, persistent problems with tissue engineering is mass transport limitations. Engineered tissues generally lack an initial blood supply, thus making it difficult for any implanted cells to obtain sufficient oxygen and nutrients to survive, and/or function properly.

Self-assembly

Self-assembly may play an important role here, both from the perspective of encapsulating cells and proteins, as well as creating scaffolds on the right physical scale for engineered tissue constructs and cellular ingrowth. The micromasonry is a prime technology to get cells grown in a lab to assemble into three-dimensional shapes. To break down tissue into single-cell building blocks, researchers have to dissolve the extracellular mortar that normally binds them together. But once that glue is removed, it's quite difficult to get cells to reassemble into the complex structures that make up our natural tissues. While cells aren't easily stackable, building blocks are. So the micromasonry starts with the encapsulation of living cells in polymer cubes. From there, the blocks self-assemble in any shape using templates.

Liquid-based Template Assembly

The air-liquid surface established by Faraday waves is explored as a template to assemble biological entities for bottom-up tissue engineering. This liquid-based template can be dynamically reconfigured in a few seconds, and the assembly on the template can be achieved in a scalable and parallel manner. Assembly of microscale hydrogels, cells, neuron-seeded micro-carrier beads, cell spheroids into various symmetrical and periodic structures was demonstrated with good cell viability. Formation of 3D neural network was achieved after 14-day tissue culture.

Additive Manufacturing

It might be possible to print organs, or possibly entire organisms using additive manufacturing techniques. A recent innovative method of construction uses an ink-jet mechanism to print precise layers of cells in a matrix of thermoreversible gel. Endothelial cells, the cells that line blood vessels, have been printed in a set of stacked rings. When incubated, these fused into a tube.

The field of three-dimensional and highly accurate models of biological systems is pioneered by multiple projects and technologies including a rapid method for creating tissues and even whole organs involves a 3D printer that can print the scaffolding and cells layer by layer into a working tissue sample or organ. The device is presented in a TED talk by Dr. Anthony Atala, M.D. the Director of the Wake Forest Institute for Regenerative Medicine, and the W.H. Boyce Professor and Chair of the Department of Urology at Wake Forest University, in which a kidney is printed on stage during the seminar and then presented to the crowd. It is anticipated that this technology will enable the production of livers in the future for transplantation and theoretically for toxicology and other biological studies as well.

Recently Multi-Photon Processing (MPP) was employed for in vivo expperiments by engineering artificial cartilage constructs. An ex vivo histological examination showed that certain pore geometry and the pre-growing of chondrocytes (Cho) prior to implantation significantly improves the performance of the created 3D scaffolds. The achieved biocompatibility was comparable to the commercially available collagen membranes. The successful outcome of this study supports the idea that hexagonal-pore-shaped hybrid organic-inorganic microstructured scaffolds in combination with Cho seeding may be successfully implemented for cartilage tissue engineering.

Scaffolding

In 2013, using a 3-d scaffolding of Matrigel in various configurations, substantial pancreatic organoids was produced in vitro. Clusters of small numbers of cells proliferated into 40,000 cells within one week. The clusters transform into cells that make either digestive enzymes or hormones like insulin, self-organizing into branched pancreatic organoids that resemble the pancreas.

The cells are sensitive to the environment, such as gel stiffness and contact with other cells. Individual cells do not thrive; a minimum of four proximate cells was required for subsequent organoid development. Modifications to the medium composition produced either hollow spheres mainly composed of pancreatic progenitors, or complex organoids that spontaneously undergo pancreatic morphogenesis and differentiation. Maintenance and expansion of pancreatic progenitors require active Notch and FGF signaling, recapitulating in vivo niche signaling interactions.

The organoids were seen as potentially offering mini-organs for drug testing and for spare insulin-producing cells.

Tissue Culture

In many cases, creation of functional tissues and biological structures in vitro requires extensive culturing to promote survival, growth and inducement of functionality. In general, the basic requirements of cells must be maintained in culture, which include oxygen, pH, humidity, temperature, nutrients and osmotic pressure maintenance.

Tissue engineered cultures also present additional problems in maintaining culture conditions. In standard cell culture, diffusion is often the sole means of nutrient and metabolite transport. However, as a culture becomes larger and more complex, such as the case with engineered organs and whole tissues, other mechanisms must be employed to maintain the culture, such as the creation of capillary networks within the tissue.

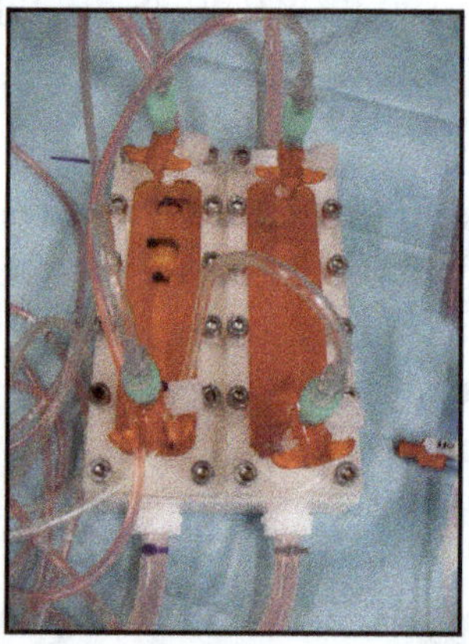

Bioreactor for cultivation of vascular grafts.

Another issue with tissue culture is introducing the proper factors or stimuli required to induce functionality. In many cases, simple maintenance culture is not sufficient. Growth factors, hormones, specific metabolites or nutrients, chemical and physical stimuli are sometimes required. For example, certain cells respond to changes in oxygen tension as part of their normal development, such as chondrocytes, which must adapt to low oxygen conditions or hypoxia during skeletal development. Others, such as endothelial cells, respond to shear stress from fluid flow, which is encountered in blood vessels. Mechanical stimuli, such as pressure pulses seem to be beneficial to all kind of cardiovascular tissue such as heart valves, blood vessels or pericardium.

Bioreactors

A bioreactor in tissue engineering, as opposed to industrial bioreactors, is a device that attempts to simulate a physiological environment in order to promote cell or tissue growth in vitro. A physiological environment can consist of many different parameters such as temperature and oxygen or carbon dioxide concentration, but can extend to all kinds of biological, chemical or mechanical stimuli. Therefore, there are systems that may include the application of forces or stresses to the tissue or even of electric current in two- or three-dimensional setups.

In academic and industry research facilities, it is typical for bioreactors to be developed to replicate the specific physiological environment of the tissue being grown (e.g., flex and fluid shearing for heart tissue growth). Several general-use and application-specific bioreactors are also commercially available, and may provide static chemical stimulation or combination of chemical and mechanical stimulation.

The Bioreactors used for 3D cell cultures are small plastic cylindrical chambers with regulated internal humidity and moisture specifically engineered for the purpose of growing cells in three dimensions. The bioreactor uses bioactive synthetic materials such as polyethylene terephthalate membranes to surround the spheroid cells in an environment that maintains high levels of nutrients. They are easy to open and close, so that cell spheroids can be removed for testing, yet the chamber is able to maintain 100% humidity throughout. This humidity is important to achieve maximum cell growth and function. The bioreactor chamber is part of a larger device that rotates to ensure equal cell growth in each direction across three dimensions. MC2 Biotek has developed a bioreactor known as ProtoTissue that uses gas exchange to maintain high oxygen levels within the cell chamber; improving upon previous bioreactors, because the higher oxygen levels help the cell grow and undergo normal cell respiration.

Long Fiber Generation

In 2013, a group from the University of Tokyo developed cell laden fibers up to a meter in length and on the order of 100 μm in size. These fibers were created using a microfluidic device that forms a double coaxial laminar flow. Each 'layer' of the microfluidic device (cells seeded in ECM, a hydrogel sheath, and finally a calcium chloride solution). The seeded cells culture within the hydrogel sheath for several days, and then the sheath is removed with viable cell fibers. Various cell types were inserted into the ECM core, including myocytes, endothelial cells, nerve cell fibers, and epithelial cell fibers. This group then showed that these fibers can be woven together to fabricate tissues or organs in a mechanism similar to textile weaving. Fibrous morphologies are advantageous in that they provide an alternative to traditional scaffold design, and many organs (such as muscle) are composed of fibrous cells.

Bioartificial Organs

An artificial organ is a man-made device that is implanted or integrated into a human to replace a natural organ, for the purpose of restoring a specific function or a group of related functions so the patient may return to a normal life as soon as possible. The replaced function doesn't necessarily have to be related to life support, but often is. The ultimate goal of tissue engineering as a discipline is to allow both 'off the shelf' bioartificial organs and regeneration of injured tissue in the body. In order to successfully create bioartificial organs from a patients stem cells, researchers continue to make improvements in the generation of complex tissues by tissue engineering. For example, much research is aimed at understanding nanoscale cues present in a cell's microenvironment.

Ceramic Engineering

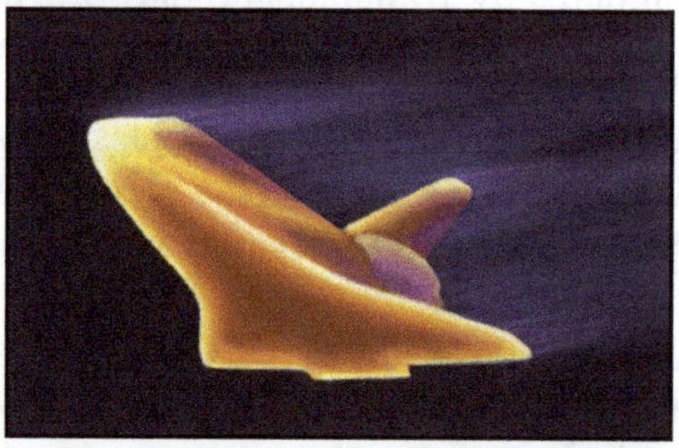

Simulation of the outside of the Space Shuttle as it heats up to over 1,500 °C (2,730 °F) during re-entry into the Earth's atmosphere.

Bearing components made from 100% silicon nitride Si_3N_4.

Ceramic bread knife.

Ceramic engineering is the science and technology of creating objects from inorganic, non-metallic materials. This is done either by the action of heat, or at lower temperatures using precipitation reactions from high-purity chemical solutions. The term includes the purification of raw materials, the study and production of the chemical compounds concerned, their formation into components and the study of their structure, composition and properties.

Ceramic materials may have a crystalline or partly crystalline structure, with long-range order on atomic scale. Glass ceramics may have an amorphous or glassy structure, with limited or short-range atomic order. They are either formed from a molten mass that solidifies on cooling, formed and matured by the action of heat, or chemically synthesized at low temperatures using, for example, hydrothermal or sol-gel synthesis.

The special character of ceramic materials gives rise to many applications in materials engineering, electrical engineering, chemical engineering and mechanical engineering. As ceramics are heat resistant, they can be used for many tasks for which materials like metal and polymers are unsuitable. Ceramic materials are used in a wide range of industries, including mining, aerospace, medicine, refinery, food and chemical industries, packaging science, electronics, industrial and transmission electricity, and guided lightwave transmission.

History

It is related to the older Indo-European language root "to burn", "Ceramic" may be used as a noun in the singular to refer to a ceramic material or the product of ceramic manufacture, or as an adjective. The plural "ceramics" may be used to refer the making of things out of ceramic materials. Ceramic engineering, like many sciences, evolved from a different discipline by today's standards. Materials science engineering is grouped with ceramics engineering to this day.

Leo Morandi's tile glazing line (circa 1945).

Abraham Darby first used coke in 1709 in Shropshire, England, to improve the yield of a smelting process. Coke is now widely used to produce carbide ceramics. Potter Josiah Wedgwood opened the first modern ceramics factory in Stoke-on-Trent, England, in 1759. Austrian chemist Carl Josef Bayer, working for the textile industry in Russia, developed a process to separate alumina from bauxite ore in 1888. The Bayer process is still used to purify alumina for the ceramic and aluminium industries. Brothers Pierre and Jacques Curie discovered piezoelectricity in Rochelle salt circa 1880. Piezoelectricity is one of the key properties of electroceramics.

E.G. Acheson heated a mixture of coke and clay in 1893, and invented carborundum, or synthetic silicon carbide. Henri Moissan also synthesized SiC and tungsten carbide in his electric arc furnace in Paris about the same time as Acheson. Karl Schröter used liquid-phase sintering to bond or "cement" Moissan's tungsten carbide particles with cobalt in 1923 in Germany. Cemented (metal-bonded) carbide edges greatly increase the durability of hardened steel cutting tools. W.H. Nernst developed cubic-stabilized zirconia in the 1920s in Berlin. This material is used as an oxygen sensor in exhaust systems. The main limitation on the use of ceramics in engineering is brittleness.

Military

Soldiers pictured during the 2003 Iraq War seen through IR transparent Night Vision Goggles.

The military requirements of World War II encouraged developments, which created a need for high-performance materials and helped speed the development of ceramic science and engineering. Throughout the 1960s and 1970s, new types of ceramics were developed in response to advances in atomic energy, electronics, communications, and space travel. The discovery of ceramic superconductors in 1986 has spurred intense research to develop superconducting ceramic parts for electronic devices, electric motors, and transportation equipment.

There is an increasing need in the military sector for high-strength, robust materials which have the capability to transmit light around the visible (0.4–0.7 micrometers) and mid-infrared (1–5 micrometers) regions of the spectrum. These materials are needed for applications requiring transparent armour. Transparent armor is a material or system of materials designed to be optically transparent, yet protect from fragmentation or ballistic impacts. The primary requirement for a transparent armour system is to not only defeat the designated threat but also provide a multi-hit capability with minimized distortion of surrounding areas. Transparent armour windows must also be compatible with night vision equipment. New materials that are thinner, lightweight, and offer better ballistic performance are being sought. Such solid-state components have found widespread use for various applications in the electro-optical field including: optical fibres for guided lightwave transmission, optical switches, laser amplifiers and lenses, hosts for solid-state lasers and optical window materials for gas lasers, and infrared (IR) heat seeking devices for missile guidance systems and IR night vision.

Modern Industry

Now a multibillion-dollar a year industry, ceramic engineering and research has established itself as an important field of science. Applications continue to expand as researchers develop new kinds of ceramics to serve different purposes.

- Zirconium dioxide ceramics are used in the manufacture of knives. The blade of the ceramic knife will stay sharp for much longer than that of a steel knife, although it is more brittle and can be snapped by dropping it on a hard surface.
- Ceramics such as alumina, boron carbide and silicon carbide have been used in bulletproof vests to repel small arms rifle fire. Such plates are known commonly as trauma plates. Similar material is used to protect cockpits of some military aircraft, because of the low weight of the material.
- Silicon nitride parts are used in ceramic ball bearings. Their higher hardness means that they are much less susceptible to wear and can offer more than triple lifetimes. They also deform less under load meaning they have less contact with the bearing retainer walls and can roll faster. In very high speed applications, heat from friction during rolling can cause problems for metal bearings; problems which are reduced by the use of ceramics. Ceramics are also more chemically resistant and can be used in wet environments where steel bearings would rust. The major drawback to using ceramics is a significantly higher cost. In many cases their electrically insulating properties may also be valuable in bearings.
- In the early 1980s, Toyota researched production of an adiabatic ceramic engine which can run at a temperature of over 6000 °F (3300 °C). Ceramic engines do not require a cooling system and hence allow a major weight reduction and therefore greater fuel efficiency. Fuel efficiency of the engine is also higher at high temperature, as shown by Carnot's theorem. In a conventional metallic engine, much of the energy released from the fuel must be dissipated as waste heat in order to prevent a meltdown of the metallic parts. Despite all of these desirable properties, such engines are not in production because the manufacturing of ceramic parts in the requisite precision and durability is difficult. Imperfection in the ceramic leads to cracks, which can lead to potentially dangerous equipment failure. Such engines are possible in laboratory settings, but mass-production is not feasible with current technology.
- Work is being done in developing ceramic parts for gas turbine engines. Currently, even blades made of advanced metal alloys used in the engines' hot section require cooling and careful limiting of operating temperatures. Turbine engines made with ceramics could operate more efficiently, giving aircraft greater range and payload for a set amount of fuel.

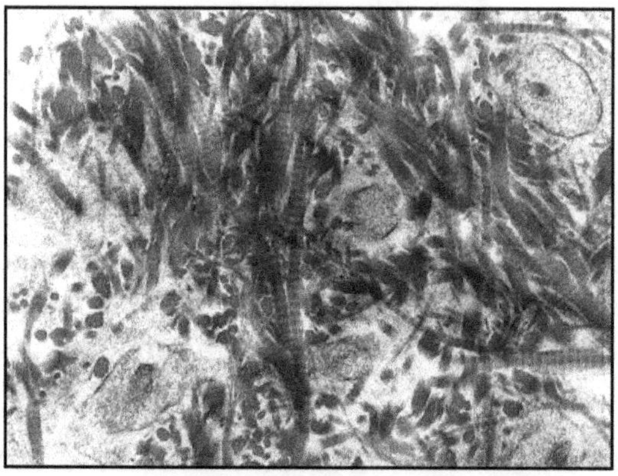

Collagen fibers of woven bone.

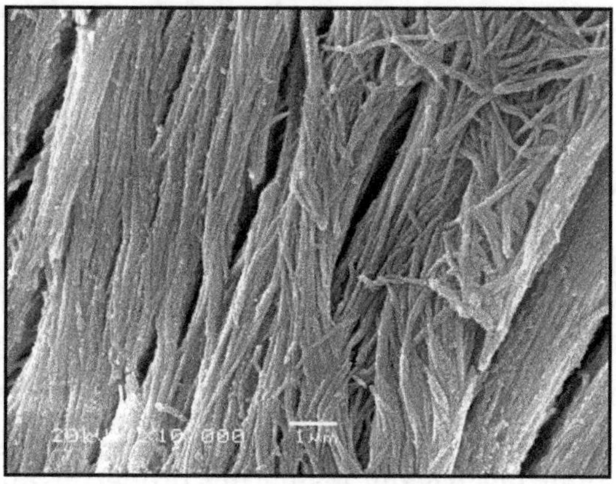

Scanning electron microscopy image of bone.

- Recently, there have been advances in ceramics which include bio-ceramics, such as dental implants and synthetic bones. Hydroxyapatite, the natural mineral component of bone, has been made synthetically from a number of biological and chemical sources and can be formed into ceramic materials. Orthopaedic implants made from these materials bond readily to bone and other tissues in the body without rejection or inflammatory reactions. Because of this, they are of great interest for gene delivery and tissue engineering scaffolds. Most hydroxyapatite ceramics are very porous and lack mechanical strength and are used to coat metal orthopaedic devices to aid in forming a bond to bone or as bone fillers. They are also used as fillers for orthopaedic plastic screws to aid in reducing the inflammation and increase absorption of these plastic materials. Work is being done to make strong, fully dense nano crystalline hydroxyapatite ceramic materials for orthopaedic weight bearing devices, replacing foreign metal and plastic orthopaedic materials with a synthetic, but naturally occurring, bone mineral. Ultimately these ceramic materials may be used as bone replacements or with the incorporation of protein collagens, synthetic bones.
- High-tech ceramic is used in watch-making for producing watch cases. The material is valued by watchmakers for its light weight, scratch-resistance, durabil-

ity and smooth touch. IWC is one of the brands that initiated the use of ceramic in watch-making. The case of the IWC 2007 Top Gun edition of the Pilot's Watch Double chronograph is crafted in high-tech black ceramic.

Glass-ceramics

A high strength glass-ceramic cook-top with negligible thermal expansion.

Glass-ceramic materials share many properties with both glasses and ceramics. Glass-ceramics have an amorphous phase and one or more crystalline phases and are produced by a so-called "controlled crystallization", which is typically avoided in glass manufacturing. Glass-ceramics often contain a crystalline phase which constitutes anywhere from 30% [m/m] to 90% [m/m] of its composition by volume, yielding an array of materials with interesting thermomechanical properties.

In the processing of glass-ceramics, molten glass is cooled down gradually before reheating and annealing. In this heat treatment the glass partly crystallizes. In many cases, so-called 'nucleation agents' are added in order to regulate and control the crystallization process. Because there is usually no pressing and sintering, glass-ceramics do not contain the volume fraction of porosity typically present in sintered ceramics.

The term mainly refers to a mix of lithium and aluminosilicates which yields an array of materials with interesting thermomechanical properties. The most commercially important of these have the distinction of being impervious to thermal shock. Thus, glass-ceramics have become extremely useful for countertop cooking. The negative thermal expansion coefficient (TEC) of the crystalline ceramic phase can be balanced with the positive TEC of the glassy phase. At a certain point (~70% crystalline) the glass-ceramic has a net TEC near zero. This type of glass-ceramic exhibits excellent mechanical properties and can sustain repeated and quick temperature changes up to 1000 °C.

Processing Steps

The traditional ceramic process generally follows this sequence: Milling → Batching → Mixing → Forming → Drying → Firing → Assembly.

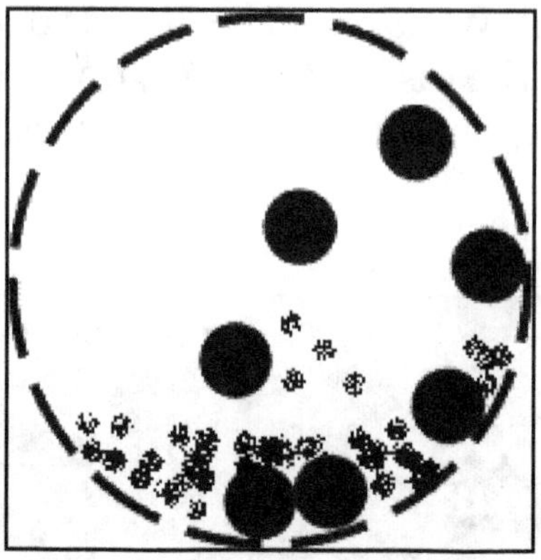

Ball mill:

- Milling is the process by which materials are reduced from a large size to a smaller size. Milling may involve breaking up cemented material (in which case individual particles retain their shape) or pulverization (which involves grinding the particles themselves to a smaller size). Milling is generally done by mechanical means, including attrition (which is particle-to-particle collision that results in agglomerate break up or particle shearing), compression (which applies a forces that results in fracturing), and impact (which employs a milling medium or the particles themselves to cause fracturing). Attrition milling equipment includes the wet scrubber (also called the planetary mill or wet attrition mill), which has paddles in water creating vortexes in which the material collides and break up. Compression mills include the jaw crusher, roller crusher and cone crusher. Impact mills include the ball mill, which has media that tumble and fracture the material. Shaft impactors cause particle-to particle attrition and compression.

- Batching is the process of weighing the oxides according to recipes, and preparing them for mixing and drying.

- Mixing occurs after batching and is performed with various machines, such as dry mixing ribbon mixers (a type of cement mixer), Mueller mixers,[clarification needed] and pug mills. Wet mixing generally involves the same equipment.

- Forming is making the mixed material into shapes, ranging from toilet bowls to spark plug insulators. Forming can involve: (1) Extrusion, such as extruding "slugs" to make bricks, (2) Pressing to make shaped parts, (3) Slip casting, as in making toilet bowls, wash basins and ornamentals like ceramic statues. Forming produces a "green" part, ready for drying. Green parts are soft, pliable, and over time will lose shape. Handling the green product will change its shape. For example, a green brick can be "squeezed", and after squeezing it will stay that way.

- Drying is removing the water or binder from the formed material. Spray drying is widely used to prepare powder for pressing operations. Other dryers are tunnel dryers and periodic dryers. Controlled heat is applied in this two-stage process. First, heat removes water. This step needs careful control, as rapid heating causes cracks and surface defects. The dried part is smaller than the green part, and is brittle, necessitating careful handling, since a small impact will cause crumbling and breaking.

- Sintering is where the dried parts pass through a controlled heating process, and the oxides are chemically changed to cause bonding and densification. The fired part will be smaller than the dried part.

Forming Methods

Ceramic forming techniques include throwing, slipcasting, tape casting, freeze-casting, injection moulding, dry pressing, isostatic pressing, hot isostatic pressing (HIP) and others. Methods for forming ceramic powders into complex shapes are desirable in many areas of technology. Such methods are required for producing advanced, high-temperature structural parts such as heat engine components and turbines. Materials other than ceramics which are used in these processes may include: wood, metal, water, plaster and epoxy—most of which will be eliminated upon firing.

These forming techniques are well known for providing tools and other components with dimensional stability, surface quality, high (near theoretical) density and microstructural uniformity. The increasing use and diversity of speciality forms of ceramics adds to the diversity of process technologies to be used.

Thus, reinforcing fibres and filaments are mainly made by polymer, sol-gel, or CVD processes, but melt processing also has applicability. The most widely used speciality form is layered structures, with tape casting for electronic substrates and packages being pre-eminent. Photo-lithography is of increasing interest for precise patterning of conductors and other components for such packaging. Tape casting or forming processes are also of increasing interest for other applications, ranging from open structures such as fuel cells to ceramic composites.

The other major layer structure is coating, where melt spraying is very important, but chemical and physical vapour deposition and chemical (e.g., sol-gel and polymer pyrolysis) methods are all seeing increased use. Besides open structures from formed tape, extruded structures, such as honeycomb catalyst supports, and highly porous structures, including various foams, for example, reticulated foam, are of increasing use.

Densification of consolidated powder bodies continues to be achieved predominantly by (pressureless) sintering. However, the use of pressure sintering by hot pressing is increasing, especially for non-oxides and parts of simple shapes where higher quality (mainly microstructural homogeneity) is needed, and larger size or multiple parts per pressing can be an advantage.

The Sintering Process

The principles of sintering-based methods are simple ("sinter" has roots in the English "cinder"). The firing is done at a temperature below the melting point of the ceramic. Once a roughly-held-together object called a "green body" is made, it is baked in a kiln, where atomic and molecular diffusion processes give rise to significant changes in the primary microstructural features. This includes the gradual elimination of porosity, which is typically accompanied by a net shrinkage and overall densification of the component. Thus, the pores in the object may close up, resulting in a denser product of significantly greater strength and fracture toughness.

Another major change in the body during the firing or sintering process will be the establishment of the polycrystalline nature of the solid. This change will introduce some form of grain size distribution, which will have a significant impact on the ultimate physical properties of the material. The grain sizes will either be associated with the initial particle size, or possibly the sizes of aggregates or particle clusters which arise during the initial stages of processing.

The ultimate microstructure (and thus the physical properties) of the final product will be limited by and subject to the form of the structural template or precursor which is created in the initial stages of chemical synthesis and physical forming. Hence the importance of chemical powder and polymer processing as it pertains to the synthesis of industrial ceramics, glasses and glass-ceramics.

There are numerous possible refinements of the sintering process. Some of the most common involve pressing the green body to give the densification a head start and reduce the sintering time needed. Sometimes organic binders such as polyvinyl alcohol are added to hold the green body together; these burn out during the firing (at 200–350 °C). Sometimes organic lubricants are added during pressing to increase densification. It is common to combine these, and add binders and lubricants to a powder, then press. (The formulation of these organic chemical additives is an art in itself. This is particularly important in the manufacture of high performance ceramics such as those used by the billions for electronics, in capacitors, inductors, sensors, etc.)

A slurry can be used in place of a powder, and then cast into a desired shape, dried and then sintered. Indeed, traditional pottery is done with this type of method, using a plastic mixture worked with the hands. If a mixture of different materials is used together in a ceramic, the sintering temperature is sometimes above the melting point of one minor component – a liquid phase sintering. This results in shorter sintering times compared to solid state sintering.

Strength of Ceramics

A material's strength is dependent on its microstructure. The engineering processes to which a material is subjected can alter its microstructure. The variety of strengthening mechanisms that alter the strength of a material include the mechanism of grain boundary strengthening. Thus, although yield strength is maximized with decreasing grain size, ultimately, very small grain sizes make the material brittle. Considered in tandem with the fact that the yield strength is the parameter that predicts plastic deformation in the material, one can make informed decisions on how to increase the strength of a material depending on its microstructural properties and the desired end effect.

The relation between yield stress and grain size is described mathematically by the Hall-Petch equation which is:

$$\sigma_y = \sigma_0 + \frac{k_y}{\sqrt{d}}$$

where k_y is the strengthening coefficient (a constant unique to each material), σ_0 is a materials constant for the starting stress for dislocation movement (or the resistance of the lattice to dislocation motion), d is the grain diameter, and σ_y is the yield stress.

Theoretically, a material could be made infinitely strong if the grains are made infinitely small. This is, unfortunately, impossible because the lower limit of grain size is a single unit cell of the material. Even then, if the grains of a material are the size of a single unit cell, then the material is in fact amorphous, not crystalline, since there is no long range order, and dislocations can not be defined in an amorphous material. It has been observed experimentally that the microstructure with the highest yield strength is a grain size of about 10 nanometres, because grains smaller than this undergo another yielding mechanism, grain boundary sliding. Producing engineering materials with this ideal grain size is difficult because of the limitations of initial particle sizes inherent to nanomaterials and nanotechnology.

Theory of Chemical Processing

Microstructural Uniformity

In the processing of fine ceramics, the irregular particle sizes and shapes in a typical powder often lead to non-uniform packing morphologies that result in packing density variations in the powder compact. Uncontrolled agglomeration of powders due to attractive van der Waals forces can also give rise to in microstructural inhomogeneities.

Differential stresses that develop as a result of non-uniform drying shrinkage are directly related to the rate at which the solvent can be removed, and thus highly dependent upon the distribution of porosity. Such stresses have been associated with a plastic-to-brittle transition in consolidated bodies, and can yield to crack propagation in the unfired body if not relieved.

In addition, any fluctuations in packing density in the compact as it is prepared for the kiln are often amplified during the sintering process, yielding inhomogeneous densification. Some pores and other structural defects associated with density variations have been shown to play a detrimental role in the sintering process by growing and thus limiting end-point densities. Differential stresses arising from inhomogeneous densification have also been shown to result in the propagation of internal cracks, thus becoming the strength-controlling flaws.

It would therefore appear desirable to process a material in such a way that it is physically uniform with regard to the distribution of components and porosity, rather than using particle size distributions which will maximize the green density. The containment of a uniformly dispersed assembly of strongly interacting particles in suspension requires total control over particle-particle interactions. Monodisperse colloids provide this potential.

Monodisperse powders of colloidal silica, for example, may therefore be stabilized sufficiently to ensure a high degree of order in the colloidal crystal or polycrystalline colloidal solid which results from aggregation. The degree of order appears to be limited by the time and space allowed for longer-range correlations to be established.

Such defective polycrystalline colloidal structures would appear to be the basic elements of submicrometer colloidal materials science, and, therefore, provide the first step in developing a more rigorous understanding of the mechanisms involved in microstructural evolution in inorganic systems such as polycrystalline ceramics.

Self-assembly

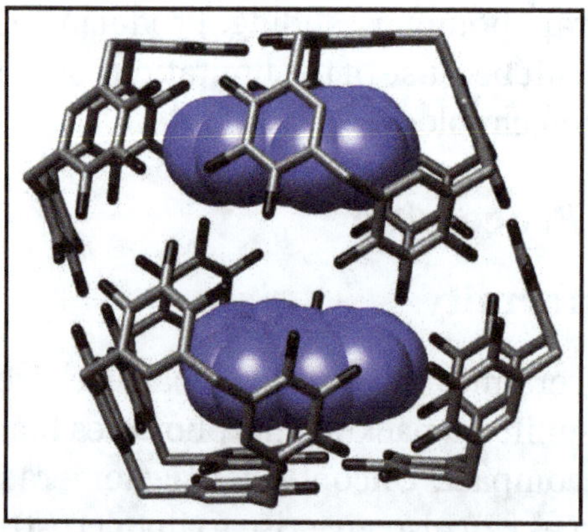

An example of a supramolecular assembly.

Self-assembly is the most common term in use in the modern scientific community to describe the spontaneous aggregation of particles (atoms, molecules, colloids, micelles, etc.) without the influence of any external forces. Large groups of such particles are known to assemble themselves into thermodynamically stable, structurally well-defined arrays, quite reminiscent of one of the 7 crystal systems found in metallurgy and mineralogy (e.g. face-centred cubic, body-centred cubic, etc.). The fundamental difference in equilibrium structure is in the spatial scale of the unit cell (or lattice parameter) in each particular case.

Thus, self-assembly is emerging as a new strategy in chemical synthesis and nanotechnology. Molecular self-assembly has been observed in various biological systems and underlies the formation of a wide variety of complex biological structures. Molecular crystals, liquid crystals, colloids, micelles, emulsions, phase-separated polymers, thin films and self-assembled monolayers all represent examples of the types of highly ordered structures which are obtained using these techniques. The distinguishing feature of these methods is self-organization in the absence of any external forces.

In addition, the principal mechanical characteristics and structures of biological ceramics, polymer composites, elastomers, and cellular materials are being re-evaluated, with an emphasis on bioinspired materials and structures. Traditional approaches focus on design methods of biological materials using conventional synthetic materials. This includes an emerging class of mechanically superior biomaterials based on microstructural features and designs found in nature. The new horizons have been identified in the synthesis of bioinspired materials through processes that are characteristic of biological systems in nature. This includes the nanoscale self-assembly of the components and the development of hierarchical structures.

Ceramic Composites

The Porsche Carrera GT's carbon-ceramic (silicon carbide) composite disc brake.

Substantial interest has arisen in recent years in fabricating ceramic composites. While there is considerable interest in composites with one or more non-ceramic constituents, the greatest attention is on composites in which all constituents are ceramic. These typically comprise two ceramic constituents: a continuous matrix, and a dispersed phase of ceramic particles, whiskers, or short (chopped) or continuous ceramic fibres. The challenge, as in wet chemical processing, is to obtain a uniform or homogeneous distribution of the dispersed particle or fibre phase.

Consider first the processing of particulate composites. The particulate phase of greatest interest is tetragonal zirconia because of the toughening that can be achieved from the phase transformation from the metastable tetragonal to the monoclinic crystalline phase, aka transformation toughening. There is also substantial interest in dispersion of hard, non-oxide phases such as SiC, TiB, TiC, boron, carbon and especially oxide matrices like alumina and mullite. There is also interest too incorporating other ceramic particulates, especially those of highly anisotropic thermal expansion. Examples include Al_2O_3, TiO_2, graphite, and boron nitride.

Silicon carbide single crystal.

In processing particulate composites, the issue is not only homogeneity of the size and spatial distribution of the dispersed and matrix phases, but also control of the matrix grain size. However, there is some built-in self-control due to inhibition of matrix grain growth by the dispersed phase. Particulate composites, though generally offer increased resistance to damage, failure, or both, are still quite sensitive to inhomogeneities of composition as well as other processing defects such as pores. Thus they need good processing to be effective.

Particulate composites have been made on a commercial basis by simply mixing powders of the two constituents. Although this approach is inherently limited in the homogeneity that can be achieved, it is the most readily adaptable for existing ceramic production technology. However, other approaches are of interest.

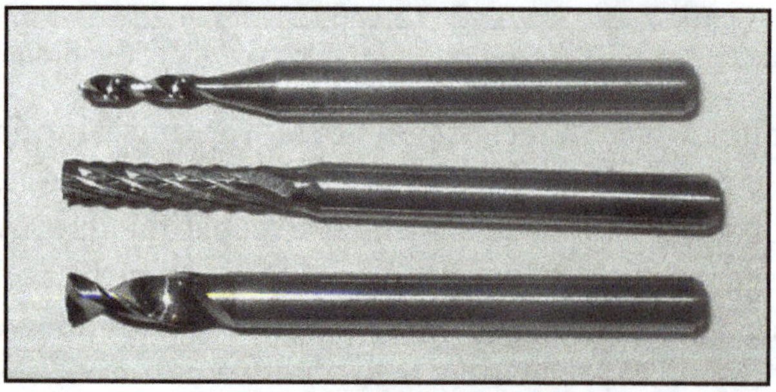

Tungsten carbide milling bits.

From the technological standpoint, a particularly desirable approach to fabricating particulate composites is to coat the matrix or its precursor onto fine particles of the dispersed phase with good control of the starting dispersed particle size and the resultant matrix coating thickness. One should in principle be able to achieve the ultimate in homogeneity of distribution and thereby optimize composite performance. This can also have other ramifications, such as allowing more useful composite performance to be achieved in a body having porosity, which might be desired for other factors, such as limiting thermal conductivity.

There are also some opportunities to utilize melt processing for fabrication of ceramic, particulate, whisker and short-fibre, and continuous-fibre composites. Clearly, both particulate and whisker composites are conceivable by solid-state precipitation after solidification of the melt. This can also be obtained in some cases by sintering, as for precipitation-toughened, partially stabilized zirconia. Similarly, it is known that one can directionally solidify ceramic eutectic mixtures and hence obtain uniaxially aligned fibre composites. Such composite processing has typically been limited to very simple shapes and thus suffers from serious economic problems due to high machining costs.

Clearly, there are possibilities of using melt casting for many of these approaches. Potentially even more desirable is using melt-derived particles. In this method, quenching is done in a solid solution or in a fine eutectic structure, in which the particles are then processed by more typical ceramic powder processing methods into a useful body. There have also been preliminary attempts to use melt spraying as a means of forming composites by introducing the dispersed particulate, whisker, or fibre phase in conjunction with the melt spraying process.

Other methods besides melt infiltration to manufacture ceramic composites with long fibre reinforcement are chemical vapour infiltration and the infiltration of fibre preforms with organic precursor, which after pyrolysis yield an amorphous ceramic matrix, initially with a low density. With repeated cycles of infiltration and pyrolysis one of those types of ceramic matrix composites is produced. Chemical vapour infiltration is used to manufacture carbon/carbon and silicon carbide reinforced with carbon or silicon carbide fibres.

Besides many process improvements, the first of two major needs for fibre composites is lower fibre costs. The second major need is fibre compositions or coatings, or composite processing, to reduce degradation that results from high-temperature composite exposure under oxidizing conditions.

Applications

The products of technical ceramics include tiles used in the Space Shuttle program, gas burner nozzles, ballistic protection, nuclear fuel uranium oxide pellets, bio-medical implants, jet engine turbine blades, and missile nose cones.

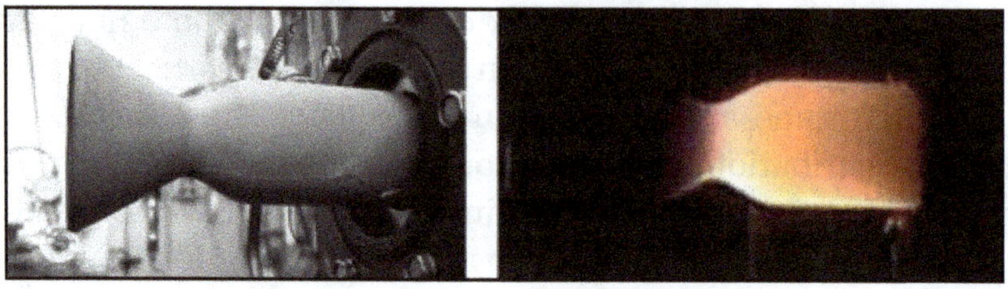

Silicon nitride thruster. Left: Mounted in test stand. Right: Being tested with H_2/O_2 propellants.

Its products are often made from materials other than clay, chosen for their particular physical properties. These may be classified as follows:

- Oxides: silica, alumina, zirconia.
- Non-oxides: carbides, borides, nitrides, silicides.
- Composites: particulate or whisker reinforced matrices, combinations of oxides and non-oxides (e.g. polymers).

Ceramics can be used in many technological industries. One application is the ceramic tiles on NASA's Space Shuttle, used to protect it and the future supersonic space planes from the searing heat of re-entry into the Earth's atmosphere. They are also used widely in electronics and optics. In addition to the applications listed here, ceramics are also used as a coating in various engineering cases. An example would be a ceramic bearing coating over a titanium frame used for an aircraft. Recently the field has come to include the studies of single crystals or glass fibres, in addition to traditional polycrystalline materials, and the applications of these have been overlapping and changing rapidly.

Aerospace

- Engines; Shielding a hot running aircraft engine from damaging other components.
- Airframes; Used as a high-stress, high-temp and lightweight bearing and structural component.
- Missile nose-cones; Shielding the missile internals from heat.
- Space Shuttle tiles.
- Space-debris ballistic shields – ceramic fiber woven shields offer better protection to hypervelocity (~7 km/s) particles than aluminium shields of equal weight.
- Rocket nozzles, withstands and focuses the exhaust of the rocket booster.
- Unmanned Air Vehicles; Implications of ceramic engine utilization in aeronautical applications (such as Unmanned Air Vehicles) may result in enhanced performance characteristics and less operational costs.

Biomedical

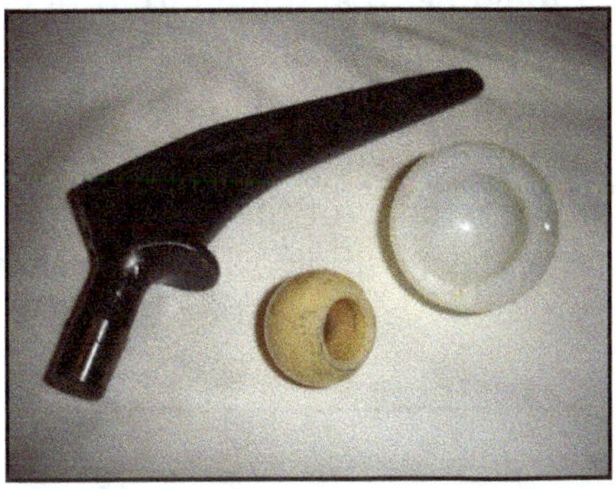

A titanium hip prosthesis, with a ceramic head and polyethylene acetabular cup.

- Artificial bone; Dentistry applications, teeth.
- Biodegradable splints; Reinforcing bones recovering from osteoporosis.
- Implant material.

Electronics

- Capacitors.
- Integrated circuit packages.
- Transducers.
- Insulators.

Optical

- Optical fibres, guided lightwave transmission.
- Switches.
- Laser amplifiers.
- Lenses.
- Infrared heat-seeking devices.

Automotive

- Heat shield.
- Exhaust heat management.

Biomaterials

Silicification is quite common in the biological world and occurs in bacteria, single-celled organisms, plants, and animals (invertebrates and vertebrates). Crystalline minerals formed in such environment often show exceptional physical properties (e.g. strength, hardness, fracture toughness) and tend to form hierarchical structures that exhibit microstructural order over a range of length or spatial scales. The minerals are crystallized from an environment that is undersaturated with respect to silicon, and under conditions of neutral pH and low temperature (0–40 °C). Formation of the mineral may occur either within or outside of the cell wall of an organism, and specific biochemical reactions for mineral deposition exist that include lipids, proteins and carbohydrates.

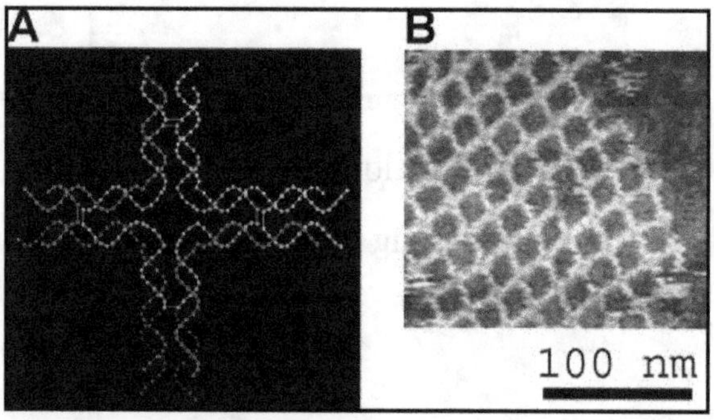

The DNA structure at left (schematic shown) will self-assemble into the structure visualized by atomic force microscopy at right.

Most natural (or biological) materials are complex composites whose mechanical properties are often outstanding, considering the weak constituents from which they are assembled. These complex structures, which have risen from hundreds of million years of evolution, are inspiring the design of novel materials with exceptional physical properties for high performance in adverse conditions. Their defining characteristics such as hierarchy, multifunctionality, and the capacity for self-healing, are currently being investigated.

The basic building blocks begin with the 20 amino acids and proceed to polypeptides, polysaccharides, and polypeptides–saccharides. These, in turn, compose the basic proteins, which are the primary constituents of the 'soft tissues' common to most biominerals. With well over 1000 proteins possible, current research emphasizes the use of collagen, chitin, keratin, and elastin. The 'hard' phases are often strengthened by crystalline minerals, which nucleate and grow in a biomediated environment that determines the size, shape and distribution of individual crystals. The most important mineral phases have been identified as hydroxyapatite, silica, and aragonite. Using the classification of Wegst and Ashby, the principal mechanical characteristics and structures of biological ceramics, polymer composites, elastomers, and cellular materials have been presented. Selected systems in each class are being investigated with emphasis on the relationship between their microstructure over a range of length scales and their mechanical response.

Thus, the crystallization of inorganic materials in nature generally occurs at ambient temperature and pressure. Yet the vital organisms through which these minerals form are capable of consistently producing extremely precise and complex structures. Understanding the processes in which living organisms control the growth of crystalline minerals such as silica could lead to significant advances in the field of materials science, and open the door to novel synthesis techniques for nanoscale composite materials, or nanocomposites.

The iridescent nacre inside a Nautilus shell.

High-resolution SEM observations were performed of the microstructure of the mother-of-pearl (or nacre) portion of the abalone shell. Those shells exhibit the highest mechanical strength and fracture toughness of any non-metallic substance known. The nacre from the shell of the abalone has become one of the more intensively studied biological structures in materials science. Clearly visible in these images are the neatly stacked (or ordered) mineral tiles separated by thin organic sheets along with a macrostructure of larger periodic growth bands which collectively form what scientists are currently referring to as a hierarchical composite structure. (The term hierarchy simply implies that there are a range of structural features which exist over a wide range of length scales).

Future developments reside in the synthesis of bio-inspired materials through processing methods and strategies that are characteristic of biological systems. These involve nanoscale self-assembly of the components and the development of hierarchical structures.

Chemometrics

Chemometrics is the science of extracting information from chemical systems by data-driven means. Chemometrics is inherently interdisciplinary, using methods frequently employed in core data-analytic disciplines such as multivariate statistics, applied mathematics, and computer science, in order to address problems in chemistry, biochemistry, medicine, biology and chemical engineering. In this way, it mirrors other interdisciplinary fields, such as psychometrics and econometrics.

Introduction

Chemometrics is applied to solve both descriptive and predictive problems in experimental natural sciences, especially in chemistry. In descriptive applications, properties of chemical systems are modeled with the intent of learning the underlying relationships and structure of the system (i.e., model understanding and identification). In predictive applications, properties of chemical systems are modeled with the intent of predicting new properties or behavior of interest. In both cases, the datasets can be small but are often very large and highly complex, involving hundreds to thousands of variables, and hundreds to thousands of cases or observations.

Chemometric techniques are particularly heavily used in analytical chemistry and metabolomics, and the development of improved chemometric methods of analysis also continues to advance the state of the art in analytical instrumentation and methodology. It is an application-driven discipline, and thus while the standard chemometric methodologies are very widely used industrially, academic groups are dedicated to the continued development of chemometric theory, method and application development.

Origins

Although one could argue that even the earliest analytical experiments in chemistry involved a form of chemometrics, the field is generally recognized to have emerged in the 1970s as computers became increasingly exploited for scientific investigation. The term 'chemometrics' was coined by Svante Wold in a grant application 1971, and the International Chemometrics Society was formed shortly thereafter by Svante Wold and Bruce Kowalski, two pioneers in the field. Wold was a professor of organic chemistry at Umeå University, Sweden, and Kowalski was a professor of analytical chemistry at University of Washington, Seattle.

Many early applications involved multivariate classification, numerous quantitative predictive applications followed, and by the late 1970s and early 1980s a wide variety of data- and computer-driven chemical analyses were occurring.

Multivariate analysis was a critical facet even in the earliest applications of chemometrics. The data resulting from infrared and UV/visible spectroscopy are often easily numbering in the thousands of measurements per sample. Mass spectrometry, nuclear magnetic resonance, atomic emission/absorption and chromatography experiments are also all by nature highly multivariate. The structure of these data was found to be conducive to using techniques such as principal components analysis (PCA), and partial least-squares (PLS). This is primarily because, while the datasets may be highly multivariate there is strong and often linear low-rank structure present. PCA and PLS have been shown over time very effective at empirically modeling the more chemically interesting low-rank structure, exploiting the interrelationships or 'latent variables' in the data, and providing alternative compact coordinate systems for further numerical analysis such as regression, clustering, and pattern recognition. Partial least squares in particular was heavily used in chemometric applications for many years before it began to find regular use in other fields.

Through the 1980s three dedicated journals appeared in the field: Journal of Chemometrics, Chemometrics and Intelligent Laboratory Systems, and Journal of Chemical Information and Modeling. These journals continue to cover both fundamental and methodological research in chemometrics. At present, most routine applications of existing chemometric methods are commonly published in application-oriented journals (e.g., Applied Spectroscopy, Analytical Chemistry, Anal. Chim. Acta., Talanta). Several important books/monographs on chemometrics were also first published in the 1980s, including the first edition of Malinowski's Factor Analysis in Chemistry, Sharaf, Illman and Kowalski's Chemometrics, Massart et al. Chemometrics: a textbook, and Multivariate Calibration by Martens and Naes.

Some large chemometric application areas have gone on to represent new domains, such as molecular modeling and QSAR, cheminformatics, the '-omics' fields of genomics, proteomics, metabonomics and metabolomics, process modeling and process analytical technology.

An account of the early history of chemometrics was published as a series of interviews by Geladi and Esbensen.

Techniques

Multivariate Calibration

Many chemical problems and applications of chemometrics involve calibration. The objective is to develop models which can be used to predict properties of interest based on measured properties of the chemical system, such as pressure, flow, temperature, infrared, Raman, NMR spectra and mass spectra. Examples include the development of multivariate models relating 1) multi-wavelength spectral response to analyte concentration, 2) molecular descriptors to biological activity, 3) multivariate process conditions/states to final product attributes. The process requires a calibration or training data set, which includes reference values for the properties of interest for prediction, and the measured attributes believed to correspond to these properties. For case 1), for example, one can assemble data from a number of samples, including concentrations for an analyte of interest for each sample (the reference) and the corresponding infrared spectrum of that sample. Multivariate calibration techniques such as partial-least squares regression, or principal component regression (and near countless other methods) are then used to construct a mathematical model that relates the multivariate response (spectrum) to the concentration of the analyte of interest, and such a model can be used to efficiently predict the concentrations of new samples.

Techniques in multivariate calibration are often broadly categorized as classical or inverse methods. The principal difference between these approaches is that in classical calibration the models are solved such that they are optimal in describing the measured analytical responses (e.g., spectra) and can therefore be considered optimal descriptors, whereas in inverse methods the models are solved to be optimal in predicting the properties of interest (e.g., concentrations, optimal predictors). Inverse methods usually require less physical knowledge of the chemical system, and at least in theory provide superior predictions in the mean-squared error sense, and hence inverse approaches tend to be more frequently applied in contemporary multivariate calibration.

The main advantages of the use of multivariate calibration techniques is that fast, cheap, or non-destructive analytical measurements (such as optical spectroscopy) can be used to estimate sample properties which would otherwise require time-consuming, expensive or destructive testing (such as LC-MS). Equally important is that multivariate calibration allows for accurate quantitative analysis in the presence of heavy interference by other analytes. The selectivity of the analytical method is provided as much by the mathematical calibration, as the analytical measurement modalities. For example near-infrared spectra, which are extremely broad and non-selective compared to other analytical techniques (such as infrared or Raman spectra), can often be used successfully in conjunction with carefully developed multivariate calibration methods to predict concentrations of analytes in very complex matrices.

Classification, Pattern Recognition, Clustering

Supervised multivariate classification techniques are closely related to multivariate calibration techniques in that a calibration or training set is used to develop a mathematical model capable of classifying future samples. The techniques employed in chemometrics are similar to those used in other fields – multivariate discriminant analysis, logistic regression, neural networks, regression/classification trees. The use of rank reduction techniques in conjunction with these conventional classification methods is routine in chemometrics, for example discriminant analysis on principal components or partial least squares scores.

Unsupervised classification (also termed cluster analysis) is also commonly used to discover patterns in complex data sets, and again many of the core techniques used in chemometrics are common to other fields such as machine learning and statistical learning.

Multivariate Curve Resolution

In chemometric parlance, multivariate curve resolution seeks to deconstruct data sets with limited or absent reference information and system knowledge. Some of the earliest work on these techniques was done by Lawton and Sylvestre in the early 1970s. These approaches are also called self-modeling mixture analysis, blind source/signal separation, and spectral unmixing. For example, from a data set comprising fluorescence spectra from a series of samples each containing multiple fluorophores, multivariate curve resolution methods can be used to extract the fluorescence spectra of the individual fluorophores, along with their relative concentrations in each of the samples, essentially unmixing the total fluorescence spectrum into the contributions from the individual components. The problem is usually ill-determined due to rotational ambiguity (many possible solutions can equivalently represent the measured data), so the application of additional constraints is common, such as non-negatively, unmodality, or known interrelationships between the individual components (e.g., kinetic or mass-balance constraints).

Other Techniques

Experimental design remains a core area of study in chemometrics and several monographs are specifically devoted to experimental design in chemical applications. Sound principles of experimental design have been widely adopted within the chemometrics community, although many complex experiments are purely observational, and there can be little control over the properties and interrelationships of the samples and sample properties.

Signal processing is also a critical component of almost all chemometric applications, particularly the use of signal pretreatments to condition data prior to calibration or classification. The techniques employed commonly in chemometrics are often closely related to those used in related fields.

Performance characterization, and figures of merit Like most arenas in the physical sciences, chemometrics is quantitatively oriented, so considerable emphasis is placed on performance characterization, model selection, verification & validation, and figures of merit. The performance of quantitative models is usually specified by root mean squared error in predicting the attribute of interest, and the performance of classifiers as a true-positive rate/false-positive rate pairs (or a full ROC curve). A recent report by Olivieri et al. provides a comprehensive overview of figures of merit and uncertainty estimation in multivariate calibration, including multivariate definitions of selectivity, sensitivity, SNR and prediction interval estimation. Chemometric model selection usually involves the use of tools such as resampling (including bootstrap, permutation, cross-validation).

Multivariate statistical process control (MSPC), modeling and optimization accounts for a substantial amount of historical chemometric development. Spectroscopy has been used successfully for online monitoring of manufacturing processes for 30–40 years, and this process data is highly amenable to chemometric modeling. Specifically in terms of MSPC, multiway modeling of batch and continuous processes is increasingly common in industry and remains an active area of research in chemometrics and chemical engineering. Process analytical chemistry as it was originally termed, or the newer term process analytical technology continues to draw heavily on chemometric methods and MSPC.

Multiway methods are heavily used in chemometric applications. These are higher-order extensions of more widely used methods. For example, while the analysis of a table (matrix, or second-order array) of data is routine in several fields, multiway methods are applied to data sets that involve 3rd, 4th, or higher-orders. Data of this type is very common in chemistry, for example a liquid-chromatography / mass spectrometry (LC-MS) system generates a large matrix of data (elution time versus m/z) for each sample analyzed. The data across multiple samples thus comprises a data cube. Batch process modeling involves data sets that have time vs. process variables vs. batch number. The multiway mathematical methods applied to these sorts of problems include PARAFAC, trilinear decomposition, and multiway PLS and PCA.

References

- Massart, D. L.; Vandeginste, B. G. M.; Deming, S. M.; Michotte, Y.; Kaufman, L. Chemometrics: a textbook. Amsterdam: Elsevier. ISBN 0444426604
- Harris, D.C., “Materials for Infrared Windows and Domes: Properties and Performance”, SPIE PRESS Monograph, Vol. PM70 (Int. Society of Optical Engineers, Bellingham WA, 2009) ISBN 978-0-8194-5978-7
- Bruns, R. E.; Scarminio, I. S.; de Barros Neto, B. (2006). Statistical design – chemometrics. Amsterdam: Elsevier. ISBN 044452181X
- Singh , R Paul; Dennis R. Heldman (2013). Introduction to Food Engineering (5th ed.). Academic Press. p. 1. ISBN 0123985307
- Franke, J. (2002). “Inverse Least Squares and Classical Least Squares Methods for Quantitative Vibrational Spectroscopy”. Handbook of Vibrational Spectroscopy. New York: Wiley. doi:10.1002/0470027320.s4603. ISBN 0471988472

Permissions

All chapters in this book are published with permission under the Creative Commons Attribution Share Alike License or equivalent. Every chapter published in this book has been scrutinized by our experts. Their significance has been extensively debated. The topics covered herein carry significant information for a comprehensive understanding. They may even be implemented as practical applications or may be referred to as a beginning point for further studies.

We would like to thank the editorial team for lending their expertise to make the book truly unique. They have played a crucial role in the development of this book. Without their invaluable contributions this book wouldn't have been possible. They have made vital efforts to compile up to date information on the varied aspects of this subject to make this book a valuable addition to the collection of many professionals and students.

This book was conceptualized with the vision of imparting up-to-date and integrated information in this field. To ensure the same, a matchless editorial board was set up. Every individual on the board went through rigorous rounds of assessment to prove their worth. After which they invested a large part of their time researching and compiling the most relevant data for our readers.

The editorial board has been involved in producing this book since its inception. They have spent rigorous hours researching and exploring the diverse topics which have resulted in the successful publishing of this book. They have passed on their knowledge of decades through this book. To expedite this challenging task, the publisher supported the team at every step. A small team of assistant editors was also appointed to further simplify the editing procedure and attain best results for the readers.

Apart from the editorial board, the designing team has also invested a significant amount of their time in understanding the subject and creating the most relevant covers. They scrutinized every image to scout for the most suitable representation of the subject and create an appropriate cover for the book.

The publishing team has been an ardent support to the editorial, designing and production team. Their endless efforts to recruit the best for this project, has resulted in the accomplishment of this book. They are a veteran in the field of academics and their pool of knowledge is as vast as their experience in printing. Their expertise and guidance has proved useful at every step. Their uncompromising quality standards have made this book an exceptional effort. Their encouragement from time to time has been an inspiration for everyone.

The publisher and the editorial board hope that this book will prove to be a valuable piece of knowledge for students, practitioners and scholars across the globe.

Index

www.ingramcontent.com/pod-product-compliance
Lightning Source LLC
LaVergne TN
LVHW080616210926
848485LV00003B/166